Mathematical Programming with Data Perturbations I

PURE AND APPLIED MATHEMATICS

A Program of Monographs, Textbooks, and Lecture Notes

Contributions to *Lecture Notes in Pure and Applied Mathematics* are reproduced by direct photography of the author's typewritten manuscript. Potential authors are advised to submit preliminary manuscripts for review purposes. After acceptance, the author is responsible for preparing the final manuscript in camera-ready form, suitable for direct reproduction. Marcel Dekker, Inc. will furnish instructions to authors and special typing paper. Sample pages are reviewed and returned with our suggestions to assure quality control and the most attractive rendering of your manuscript. The publisher will also be happy to supervise and assist in all stages of the preparation of your camera-ready manuscript.

LECTURE NOTES
IN PURE AND APPLIED MATHEMATICS

1. *N. Jacobson,* Exceptional Lie Algebras
2. *L.-Å. Lindahl and F. Poulsen,* Thin Sets in Harmonic Analysis
3. *I. Satake,* Classification Theory of Semi-Simple Algebraic Groups
4. *F. Hirzebruch, W. D. Newmann, and S. S. Koh,* Differentiable Manifolds and Quadratic Forms (out of print)
5. *I. Chavel,* Riemannian Symmetric Spaces of Rank One
6. *R. B. Burckel,* Characterization of C(X) Among Its Subalgebras
7. *B. R. McDonald, A. R. Magid, and K. C. Smith,* Ring Theory: Proceedings of the Oklahoma Conference
8. *Y.-T. Siu,* Techniques of Extension of Analytic Objects
9. *S. R. Caradus, W. E. Pfaffenberger, and B. Yood,* Calkin Algebras and Algebras of Operators on Banach Spaces
10. *E. O. Roxin, P.-T. Liu, and R. L. Sternberg,* Differential Games and Control Theory
11. *M. Orzech and C. Small,* The Brauer Group of Commutative Rings
12. *S. Thomeier,* Topology and Its Applications
13. *J. M. López and K. A. Ross,* Sidon Sets
14. *W. W. Comfort and S. Negrepontis,* Continuous Pseudometrics
15. *K. McKennon and J. M. Robertson,* Locally Convex Spaces
16. *M. Carmeli and S. Malin,* Representations of the Rotation and Lorentz Groups: An Introduction
17. *G. B. Seligman,* Rational Methods in Lie Algebras
18. *D. G. de Figueiredo,* Functional Analysis: Proceedings of the Brazilian Mathematical Society Symposium
19. *L. Cesari, R. Kannan, and J. D. Schuur,* Nonlinear Functional Analysis and Differential Equations: Proceedings of the Michigan State University Conference
20. *J. J. Schäffer,* Geometry of Spheres in Normed Spaces
21. *K. Yano and M. Kon,* Anti-Invariant Submanifolds
22. *W. V. Vasconcelos,* The Rings of Dimension Two
23. *R. E. Chandler,* Hausdorff Compactifications
24. *S. P. Franklin and B. V. S. Thomas,* Topology: Proceedings of the Memphis State University Conference
25. *S. K. Jain,* Ring Theory: Proceedings of the Ohio University Conference
26. *B. R. McDonald and R. A. Morris,* Ring Theory II: Proceedings of the Second Oklahoma Conference
27. *R. B. Mura and A. Rhemtulla,* Orderable Groups
28. *J. R. Graef,* Stability of Dynamical Systems: Theory and Applications
29. *H.-C. Wang,* Homogeneous Banach Algebras
30. *E. O. Roxin, P.-T. Liu, and R. L. Sternberg,* Differential Games and Control Theory II
31. *R. D. Porter,* Introduction to Fibre Bundles
32. *M. Altman,* Contractors and Contractor Directions Theory and Applications
33. *J. S. Golan,* Decomposition and Dimension in Module Categories
34. *G. Fairweather,* Finite Element Galerkin Methods for Differential Equations
35. *J. D. Sally,* Numbers of Generators of Ideals in Local Rings
36. *S. S. Miller,* Complex Analysis: Proceedings of the S.U.N.Y. Brockport Conference
37. *R. Gordon,* Representation Theory of Algebras: Proceedings of the Philadelphia Conference
38. *M. Goto and F. D. Grosshans,* Semisimple Lie Algebras
39. *A. I. Arruda, N. C. A. da Costa, and R. Chuaqui,* Mathematical Logic: Proceedings of the First Brazilian Conference
40. *F. Van Oystaeyen,* Ring Theory: Proceedings of the 1977 Antwerp Conference

41. *F. Van Oystaeyen and A. Verschoren,* Reflectors and Localization: Application to Sheaf Theory
42. *M. Satyanarayana,* Positively Ordered Semigroups
43. *D. L. Russell,* Mathematics of Finite-Dimensional Control Systems
44. *P.-T. Liu and E. Roxin,* Differential Games and Control Theory III: Proceedings of the Third Kingston Conference, Part A
45. *A. Geramita and J. Seberry,* Orthogonal Designs: Quadratic Forms and Hadamard Matrices
46. *J. Cigler, V. Losert, and P. Michor,* Banach Modules and Functors on Categories of Banach Spaces
47. *P.-T. Liu and J. G. Sutinen,* Control Theory in Mathematical Economics: Proceedings of the Third Kingston Conference, Part B
48. *C. Byrnes,* Partial Differential Equations and Geometry
49. *G. Klambauer,* Problems and Propositions in Analysis
50. *J. Knopfmacher,* Analytic Arithmetic of Algebraic Function Fields
51. *F. Van Oystaeyen,* Ring Theory: Proceedings of the 1978 Antwerp Conference
52. *B. Kedem,* Binary Time Series
53. *J. Barros-Neto and R. A. Artino,* Hypoelliptic Boundary-Value Problems
54. *R. L. Sternberg, A. J. Kalinowski, and J. S. Papadakis,* Nonlinear Partial Differential Equations in Engineering and Applied Science
55. *B. R. McDonald,* Ring Theory and Algebra III: Proceedings of The Third Oklahoma Conference
56. *J. S. Golan,* Structure Sheaves over a Noncommutative Ring
57. *T. V. Narayana, J. G. Williams, and R. M. Mathsen,* Combinatorics, Representation Theory and Statistical Methods in Groups: YOUNG DAY Proceedings
58. *T. A. Burton,* Modeling and Differential Equations in Biology
59. *K. H. Kim and F. W. Roush,* Introduction to Mathematical Consensus Theory
60. *J. Banas and K. Goebel,* Measures of Noncompactness in Banach Spaces
61. *O. A. Nielsen,* Direct Integral Theory
62. *J. E. Smith, G. O. Kenny, and R. N. Ball,* Ordered Groups: Proceedings of the Boise State Conference
63. *J. Cronin,* Mathematics of Cell Electrophysiology
64. *J. W. Brewer,* Power Series Over Commutative Rings
65. *P. K. Kamthan and M. Gupta,* Sequence Spaces and Series
66. *T. G. McLaughlin,* Regressive Sets and the Theory of Isols (in press)
67. *T. L. Herdman, S. M. Rankin, III, and H. W. Stech,* Integral and Functional Differential Equations
68. *R. Draper,* Commutative Algebra: Analytic Methods (in press)
69. *W. G. McKay and J. Patera,* Tables of Dimensions, Indices, and Branching Rules for Representations of Simple Lie Algebras
70. *R. L. Devaney and Z. H. Nitecki,* Classical Mechanics and Dynamical Systems
71. *J. Van Geel,* Places and Valuations in Noncommutative Ring Theory
72. *C. Faith,* Injective Modules and Injective Quotient Rings
73. *A. Fiacco,* Mathematical Programming with Data Perturbations I
74. *P. Schultz, C. Praeger, and R. Sullivan,* Algebraic Structures and Applications: Proceedings of the First Western Australian Conference on Algebra
75. *L. Bican, T. Kepka, and P. Nemec,* Rings, Modules, and Preradicals

Other Volumes in Preparation

Mathematical Programming with Data Perturbations I

Edited by Anthony V. Fiacco
The George Washington University
Washington, D.C.

MARCEL DEKKER, INC. *New York and Basel*

Library of Congress Cataloging in Publication Data
Main entry under title:

Mathematical programming with data perturbations I.

(Lecture notes in pure and applied mathematics ; 73)
"Papers presented to the First Symposium on Mathematical Programming with Data Perturbations, held on May 24-25, 1979, at the George Washington University"--Pref.
Includes index.
1. Programming (Mathematics)--Addresses, essays, lectures. 2. Perturbation (Mathematics)--Addresses, essays, lectures. I. Fiacco, Anthony V. II. Symposium on Mathematical Programming with Data Perturbations (1st : 1979 : George Washington University) III. Series.
QA402.5.M355 519.7 81-17472
ISBN 0-8247-1543-8 AACR2

MARCEL DEKKER, INC.
270 Madison Avenue, New York, New York 10016

Current printing (last digit):
10 9 8 7 6 5 4 3 2 1

PRINTED IN THE UNITED STATES OF AMERICA

PREFACE

The articles in this collection are based on papers presented to the first Symposium on Mathematical Programming with Data Perturbations, held on May 24-25, 1979 at The George Washington University. Fifteen papers in diverse areas of theory and applications were presented and approximately 50 people attended from several countries. The meeting, organized by Steve Robinson and me, was a gratifying experience and an important milestone without precedent. It was the first conference devoted entirely to sensitivity and stability analysis results, applications and questions, relative to general--including linear, nonlinear, integer, and stochastic--classes of mathematical programming problems. It confirmed the interest in the subject and the fact that the field had achieved a respectable level of maturity. Indeed it had.

Anyone following developments in this area knows that there has been an explosion of activity in the last decade. Basic existence theorems and characterizations relating solution changes to perturbations in the problem data have been known for some time for general classes of mathematical programs. But a systematic development of deep results tailored to the precise structure of important classes of problems was not available until recently. There are still many gaps in the theory that need to be filled, but the fabric that has been woven is already rich enough to conclude that every aspect of mathematical programming will be touched by the results.

The wide applicability of sensitivity and stability analysis is not surprising, since this is an essential part of any complete scientific or mathematical analysis, is universally applicable, and is as old as mathematics. Hence, it was only a matter of time that the intricacies of the supporting theory were developed for nonlinear programming. What is astonishing is that it took so long to attract the concentrated attention that was required. There continues to be a neglect of its use in the mainstream of both theoretical and practical applications.

The motivation for this volume, and for the symposium on which it is based, is to contribute to the unification of a sensitivity and stability analysis methodology. As the existing sophisticated theoretical and computational techniques are molded into a coherent body of results, they will be more accessible and visible and it is hoped that their use will become widespread. We are aware of at least two crucial additional requirements. First, the results must be made intelligible to the entire community of mathematical programmers, operations researchers, and any potentially interested user of these methods, not just to the mathematician who has specialized in this area. Second, for the practitioners, the theoretical findings must be forged into algorithmic methods that are readily understandable and computable, and for methodologists, practical applications and problems must be identified and articulated in a manner that does not presuppose expertise in the given area of application. One way to stimulate and accelerate this process is to encourage a dialogue between theoreticians and practitioners. The symposium is one vehicle for doing this, and also provides an open forum that allows for the introduction of new applications and problems and often novel and tailored techniques for calculating sensitivity and stability information. This dialogue introduces possible interfaces with other "unconventional" approaches and contexts that might not otherwise be envisaged and is reflected by some of the contributions herein.

The richness of the theory and the variety of the applications are abundantly evident in this collection of papers, which we have roughly categorized into two types: those providing basic theoretical results that are applicable to large classes of problems and those either focusing on a specific kind of application or pointing to new problem areas and techniques, to possibly new interfaces with the mainstream of existing methodology. The reader will find unified treatments of generic conditions, relationships between various important types of convergence, differentiable stability, elements of convex analysis and extensions to infinite dimensional spaces, and new approaches to stability characterizations, in the first collection. The second collection presents parametric analysis techniques for geometric programming, applications of sensitivity analysis information to cost benefit analysis, novel approaches to systematizing stability techniques for combinatorial problems, purposeful data perturbation techniques and error analysis for facilitating the storage and manipulation of large data banks, stability analysis for important applications using continuation methods, and the synthesis of the Kalman filter and dynamic programming techniques for estimating parameters in an intricate stochastic process developed for an extremely important area of application. Applications areas include stream water pollution abatement, groundwater basin development, transient stability of power systems, and stability of aircraft.

A unified framework for much of the theory reported in Part I is already in evidence, though many gaps remain to be filled. In Part II, though several applications are clearly based on "mainstream" theoretical results, important uses of sensitivity and stability analysis are demonstrated and new uses are suggested. Other applications are based on novel, tailored, often ad hoc techniques, whose extensions, connections, and possible interfaces with the mainstream body of results remain to be explored. Many new directions of research are indicated.

In closing, the editor expresses his gratitude to the contributors of this volume, to the referees whose careful and timely reviews have led to numerous significant improvements, to The George Washington University, particularly the School of Engineering and Applied Science, the Department of Operations Research, and the Institute for Management Science and Engineering, for sponsoring and strongly supporting the symposium on which this volume is based.

Anthony V. Fiacco

CONTENTS

CONTRIBUTORS

ADI BEN-ISRAEL Department of Mathematical Sciences, University of Delaware, Newark, Delaware

JOHN J. DINKEL* Division of Management Science, The Pennsylvania State University, State College, Pennsylvania

ANTHONY V. FIACCO Department of Operations Research, School of Engineering and Applied Science, The George Washington University, Washington, D.C.

R. GARDNER† Department of Mathematical Sciences, University of Petroleum and Minerals, Dhahran, Saudi Arabia

JACQUES GAUVIN Département de Mathématiques Appliquées, Ecole Polytechnique de Montréal, Montréal, Québec, Canada

ABOLFAZL GHAEMI‡ Department of Operations Research, School of Engineering and Applied Science, The George Washington University, Washington, D.C.

STEPHEN C. GRAVES Operations Research Center and Sloan School of Management, Massachusetts Institute of Technology, Cambridge, Massachusetts

HARVEY J. GREENBERG Energy Information Administration, U.S. Department of Energy, Washington, D.C.

DUANE R. HAMPTON Department of Civil Engineering, Colorado State University, Fort Collins, Colorado

WILLIAM P. HUTZLER The Rand Corporation, Washington, D.C.

GARY A. KOCHENBERGER Department of Management Science, College of Business Administration, The Pennsylvania State University, University Park, Pennsylvania

JOHN W. LABADIE Department of Civil Engineering, Colorado State University, Fort Collins, Colorado

HELMUT MAURER Institut für Numerische und Instrumentelle Mathematik, Westfälische Wilhelms-Universität, Münster, Federal Republic of Germany

---

*Current affiliation: Department of Business Analysis and Research, Texas A&M University, College Station, Texas

†Affiliation at time of writing: Department of Mathematics, Auburn University, Auburn, Alabama

‡Current affiliation: Reservoir Engineering Department, Iranian Marine International Oil Company, Teheran, Iran

RAMAN K. MEHRA Scientific Systems, Inc., Cambridge, Massachusetts

RICHARD P. O'NEILL Energy Information Administration, U.S. Department of Energy, Washington, D.C.

JEREMY F. SHAPIRO Operations Research Center, Massachusetts Institute of Technology, Cambridge, Massachusetts

JONATHAN E. SPINGARN School of Mathematics, Georgia Institute of Technology, Atlanta, Georgia

ROBERT B. WASHBURN, JR. Scientific Systems, Inc., Cambridge, Massachusetts

ROGER J. B. WETS Department of Mathematics, University of Kentucky, Lexington, Kentucky

DANNY S. WONG Faculty of Management Sciences, The Ohio State University, Columbus, Ohio

SANJO ZLOBEC Department of Mathematics, McGill University, Montreal, Quebec, Canada

Mathematical Programming with Data Perturbations I

Part I BASIC RESULTS

Chapter 1 SECOND-ORDER CONDITIONS THAT ARE NECESSARY WITH PROBABILITY ONE*

JONATHAN E. SPINGARN / Georgia Institute of Technology, Atlanta, Georgia

I. INTRODUCTION

In nonlinear programming theory there is a large gap between the weak first-order conditions that are necessary for optimality and the much stronger second-order conditions that have been found useful in the design and analysis of algorithms. It is common practice to assume (without giving any real mathematical justification) that very strong optimality conditions are satisfied at a minimizer, and to base convergence proofs, and thus to justify algorithms, on the basis of such assumptions. Of course, for any given problem, those *a priori* assumptions cannot be checked, unless the solution is already known.

In this paper, we discuss a "generic" approach to optimality conditions that has been developed in Spingarn and Rockafellar [10] and Spingarn [7,8,9]. Rather than talking about conditions that are necessary for optimality in *specific* problems, we discuss instead conditions necessary for optimality for *most* problems in a *family* of problems. More

---

*This research was supported, in part, by the Air Force Office of Scientific Research, under grant number 79-0120 at the Georgia Institute of Technology, Atlanta.

precisely, for a family (Q(p)) of nonlinear programming problems indexed by a parameter $p \in P \subset R^n$ we study conditions which, unless p belongs to a negligible set, hold at all local minimizers for (Q(p)) where by *negligible* we mean a first category set of measure zero in P.

This approach gives a rigorous mathematical underpinning to the *a priori* assumption of conditions which are not truly necessary for optimality, by describing the exact sense and the circumstances in which these conditions can be expected to hold. Another attractive feature of the theory is that "constraint qualifications," which are normally required to prove the necessity of Kuhn-Tucker type first-order conditions, need not be assumed to obtain conditions which are merely *generically* necessary.

In this paper, no proofs are presented. Instead, we refer the reader to the references [7,8,10].

II. A SIMPLE CLASS OF PERTURBATIONS

Consider the basic problem

$$\begin{aligned}&\min \; f(x) \text{ over all } x \in R^n \\ &\text{such that } g(x) \leqq 0 \text{ and } h(x) = 0\end{aligned} \tag{Q}$$

where the functions $f : R^n \to R$, $g : R^n \to R^m$, and $h : R^n \to R^k$ are continuously differentiable.

The standard *first-order conditions* for local optimality of x in (Q) are that x should be feasible and there should exist vectors $y \in R^m_+$ and $z \in R^k$ such that

$$\begin{aligned}&\nabla f(x) + y'\nabla g(x) + z'\nabla h(x) = 0 \\ &\text{and for all } i \notin I_+(x) \qquad y_i = 0\end{aligned} \tag{KT}$$

where

$$I_+(x) = \{i : 1 \leqq i \leqq m,\ g_i(x) = 0\}$$

These conditions are not actually *necessary* for optimality. They are only necessary under an additional assumption called a "constraint qualification," the simplest such being

$$\begin{aligned}&\{\nabla g_i(x) : i \in I_+(x)\} \cup \{\nabla h_j(x) : j = 1,\ldots,k\} \\ &\text{is linearly independent}\end{aligned} \tag{CQ}$$

When the functions f, g, and h are twice differentiable, a vector x is said to satisfy the *strong second-order conditions* for local optimality in (Q) if (CQ) holds, and there exist $y \in R_+^m$ and $z \in R^k$ such that (KT) holds with $y_i > 0$ for all $i \in I_+(x)$, and every nonzero $w \in R^n$ for which $w \cdot \nabla g_i(z) = 0$, for all $i \in I_+(x)$, and $w \cdot \nabla h_j(x) = 0$ for all j also satisfies $w \cdot H(x,y,z)w > 0$, where H(x,y,z) is the Hessian of the Lagrangian function in (Q):

$$H(x,y,z) = \nabla^2 f(x) + \sum_{i=1}^{m} y_i \nabla^2 g_i(x) + \sum_{j=1}^{k} z_j \nabla^2 h_j(x)$$

These conditions are known to guarantee that x is an isolated locally optimal solution to (Q). They also have other important consequences, for example with respect to the sensitivity of x to changes in a parameter; cf. Hestenes [3], Fiacco [1]. The strong conditions are useful for proving convergence results; for example, cf. Robinson [5], Rockafellar [6], Powell [4], Fiacco and McCormick [2].

Let us embed (Q) in the following family of nonlinear programming problems

$$\begin{aligned} &\min \quad f(x) - x \cdot v \text{ over all } x \in R^n \\ &\text{such that } g(x) \leqq u,\ h(x) = t \end{aligned} \qquad (Q(v,u,t))$$

The original problem (Q) then coincides with Q(0,0,0). Any particular problem in this family may be "bad" in the sense that the strong conditions may fail to hold at some local minimizer for that problem. However the set of bad problems is small, as the following shows [10]:

THEOREM 1 *Suppose* f *is of class* C^2 *and* g *and* h *are of class* C^{n-k}. *Then except for* (v,u,t) *belonging to a set of measure zero in* $R^n \times R^m \times R^k$, (Q(v,u,t)) *is such that every local optimal solution* x *satisfies the strong second-order conditions.*

III. GENERAL PERTURBATIONS

Next, we examine what happens when more general families of problems are allowed. The families we wish to consider are of the form

$$\min \quad f(x,p) \text{ over all } x \text{ satisfying } g(x,p) \leqq 0,\ h(x,p) = 0 \qquad (Q(p))$$

with p ranging over some open subset P of Euclidean space. The family Q(v,u,t) just considered clearly is a special case.

Obviously, some additional assumption is required in order to guarantee that the strong conditions fail only in a negligible subfamily. After all, we could start with a "bad" problem (Q) for which the strong conditions fail at some local minimum, and then, by introducing trivial perturbations so that (Q(p)) = (Q) for all p, we would obtain a family for which the conditions fail for every problem. The problem here is that the indicated family would not be "rich" enough; it would not contain enough perturbations.

The following definitions specify two different ways a family can be "rich." If g and h are of class C^1, let us say that the family (Q(p) is *full with respect to constraints* if the Jacobian of the function $p \to (g(x,p),h(x,p)) \in R^{m+k}$ has full rank m + k at every $(x,p) \in R^n \times P$. For any $w = (x,y,z) \in R^r$ $(r = n + m + k)$ and $p \in P$, let

$$L(w,p) = f(x,p) + y'g(x,p) + z'h(x,p)$$

be the Lagrangian for (Q(p)). If f, g, and h are of class C^2, the family (Q(p)) will be called *full* provided the function $p' \to \nabla_w L(w,p') \in R^r$ has full rank r at all $(w,p) \in R^r \times P$. Every full family is automatically full with respect to constraints. These two properties are sufficient to guarantee the generic necessity of the first-order (KT) and strong second-order conditions, respectively:

THEOREM 2 *(a) Let* g *and* h *be of class* C^s *on* $R^n \times P$ *with* $s > \max(0, n-k)$ *and let* (Q(p)) *be full with respect to constraints. Then there is a subset* $P' \subset P$ *with negligible complement such that if* $\bar{p} \in P'$ *and* $\bar{x}$ *is a local minimizer for* $(Q(\bar{p}))$*, then there exists* $(\bar{y},\bar{z}) \in R^m_+ \times R^k$ *satisfying* (KT).

(b) Let f *be of class* C^2 *and* g *and* h *of class* C^s *on* $R^n \times P$ *with* $s > \max(1, n - k)$. *If* (Q(p)) *is full, then there is a subset* $P' \subset P$ *with negligible complement such that for all* $\bar{p} \in P'$*: if* x *is a local minimizer for* $(Q(\bar{p}))$ *there exists* $(\bar{y},\bar{z}) \in R^m_+ \times R^k$ *satisfying the strong second-order conditions.*

To see how Theorem 2 can be applied, consider again the family (Q(v,u,t)). We take p = (v,u,t), so for any w = (x,y,z),

$$L(w,p) = f(x) - x \cdot v + y'(g(x) - u) + z'(h(x) - t)$$

We may then compute

$$\nabla_w L(w,p) = \begin{pmatrix} \nabla f(x) - v + \sum y_i \nabla g_i(x) + \sum z_j \nabla h_j(x) \\ \vdots \\ g_i(x) - u_i \\ \vdots \\ h_j(x) - t_j \\ \vdots \end{pmatrix}$$

and hence $\nabla_p \nabla_w L(w,p) = -I$, where I is the $(n + m + k)$-dimensional identity matrix, which is trivially of rank $n + m + k$.

The full rank criteria given in Theorem 2 are sufficient, but not necessary for the generic necessity of the strong conditions. However, the rank criteria can be weakened (and thus the theorem strengthened) slightly. To illustrate, consider the family

$$\text{minimize} \quad x^4 + p^2 x \text{ over all } x \in R \qquad (Q(p))$$

The Lagrangian for $(Q(p))$ is $L(x,p) = x^4 + p^2 x$ (since there are no constraints) so $\nabla_p \nabla_x L(x,p) = 2p$. For Theorem 2 to apply, it would have to be true that $2p \neq 0$ for all p. This is not a real obstacle though, since the theorem could be applied to the subfamily $\{Q(p) : p \neq 0\}$. The same reasoning shows in general that the result of the theorem holds whenever the set of p values for which the rank condition fails is contained in a closed measure zero subset of P:

COROLLARY 1 *If there is a closed subset $P' \subset P$ of measure zero such that the subfamily $\{(Q(p)) : p \in P \setminus P'\}$ is full (with respect to constraints), then the conclusion of Theorem 2a (resp., of Theorem 2b) holds.*

Another minor extension is suggested by the family

$$\text{minimize} \quad px^2 + (1 - p)x \text{ over all } x \in R \qquad (Q(p))$$

where $p \in R$. In this case, $\nabla_p \nabla_x L(x,p) = 2x - 1$. For Theorem 1 to apply, it would have to be the case that $2x - 1 \neq 0$ for all $x \in R$. Nonetheless, it is possible to conclude in such an instance that except for p in a negligible set, the strong conditions hold for (Q(p)) at all local minimizers other than possibly $x = \frac{1}{2}$:

COROLLARY 2 *If there is a closed set* $K \in R^n$ *such that the rank condition of Theorem 2 holds except for* $x \in K$, *then the conclusion of that theorem holds, except possibly at minimizers which are in* K.

IV. FAMILIES WITH SELECTIVE PERTURBATIONS

We are confronted with additional questions when we consider a family like the following one:

$$\begin{aligned} &\min \quad f(x) - x \cdot v \text{ over } x \in R^n \\ &\text{s.t.} \quad g(x) \leqq u,\ h(x) = t, \text{ and } x \geqq 0 \end{aligned} \qquad (S(v,u,t))$$

This family is identical to Q(v,u,t), with the important exception that here there is an additional "fixed" constraint $x \geqq 0$ that is independent of the parameters. Neither Theorem 1 nor 2 can be applied in this situation.

Those theorems *would* apply, were we to alter the family by replacing the fixed constraint with a perturbed constraint $x \geqq s$. This would yield a family Q(v,u,t,s) for which the strong conditions are necessary except for (v,u,t,s) in a set of measure zero. However, the family of interest, namely (S(v,u,t)) = (Q(v,u,t,0)), would be a measure zero subfamily of (Q(v,u,t,s)). Thus, although the set of "bad" problems in (Q(v,u,t,s)) is negligible, it does not follow that the bad problems in S(v,u,t) are negligible *with respect to* S(v,u,t).

Rather than concentrate on this particular family, we study the generic behavior of more general families of the form

$$\begin{aligned} &\min \quad f(x,p) \text{ over all } x \in R^n \\ &\text{s.t.} \quad g(x,p) \leqq 0,\ h(x,p) = 0, \text{ and } x \in C \end{aligned} \qquad (S(p))$$

where C is a fixed set. For the family S(v,u,t), we would take $C = R^n_+$, while the situation in Theorems 1 and 2 requires $C = R^n$. Concerning the family (S(p)), we will address ourselves here to three questions:

(1) What reasonable assumptions can we impose on the set C which allow us to develop a theory of generic second-order conditions for (S(p))? Intuition suggests that C must be "piecewise C^2-smooth" in some sense.

(2) What are the appropriate generic second-order conditions? One might expect that the generic conditions for S(p) would be the same as the usual strong conditions for the (equivalent) problem in which the set C is replaced by additional equality and inequality constraints. Sometimes this is true; for instance, the generic conditions for S(v,u,t) are the same as the usual strong conditions for the problem Q(v,u,t,0). But it is not always true, and in general the generic conditions depend on the set C.
(3) What "rank condition" ensures that these conditions are generic for (S(p))?

We begin by stating our assumptions on the set C. These have been incorporated into the definition of "cyrtohedron." The name is taken from the Greek "κυρτοσ" (= curved, bent) + "εδρα" (= side), and is motivated by the fact that these sets look like polyhedra, except that the "faces," instead of being polyhedral, are submanifolds.

Let $U \subset R^n$ be an open set, G_α, $\alpha \in A$ and H_β, $\beta \in B$, finite collections of differentiable functions on U. For any $A_0 \subset A$ and $x \in U$, define

$$\Gamma(x,A_0) = \{\nabla G_\alpha(x) : \alpha \in A_0\} \cup \{\nabla H_\beta(x) : \beta \in B\}$$

$$Z(A_0) = \{y \in U : 0 = G_\alpha(y) = H_\beta(y) \ \forall\, \alpha \in A_0, \ \forall\, \beta \in B\}$$

A nonempty connected set $C \subset R^n$ is a *cyrtohedron* of class C^s $(s \geqq 1)$ if for every $\bar{x} \in C$, there are finitely many C^s functions G_α, $\alpha \in A$, and H_β, $\beta \in B$, defined on a neighborhood $U \subset R^n$ of $\bar{x}$ such that $\bar{x} \in Z(A)$ and

(a) For all $x \in U$, $x \in C$ if, and only if, $G_\alpha(x) \leqq 0$ for all $\alpha \in A$ and $H_\beta(x) = 0$ for all $\beta \in B$.

(b) If $\sum_A a_\alpha \nabla G_\alpha(\bar{x}) + \sum_B b_\beta \nabla H_\beta(\bar{x}) = 0$ for some $a \in R^A_+$ and $b \in R^B$, then $a = 0$ and $b = 0$.

(c) For each $A_0 \subset A$ there is an integer $s(A_0)$ such that rank $\Gamma(x,A_0) = s(A_0)$ for all $x \in U$.

Some *examples of cyrtohedra* are (a) *a differentiable submanifold* in R^n is a cyrtohedron for which the set A may always be taken to be empty; (b) cyrtohedra for which the set A may always be taken either empty or of cardinality one are *submanifolds with boundary*; (c) a *polyhedral convex set* is the intersection of a finite number of closed half-spaces in R^n; (d) sets that can be expressed as $C = \{x \in R^n : g_i(x) \leqq 0,\ i = 1,\ldots,m,$ and $h_j(x) = 0,\ j = 1,\ldots,p\}$, where the functions g_i and h_j are of class C^k and have the property that for every $x \in C$, $\{\nabla g_i(x) : i \in I_+(x)\} \cup$

$\{\nabla h_j(x) : j = 1,\ldots,p\}$ is linearly independent, where $I_+(x) = \{i : g_i(x) = 0\}$.

For an example of a simple set that is not a cyrtohedron, consider the set $C \subset R^3$ which consists of all $x = (x_1,x_2,x_3)$ such that $|x| \leq 1$, $x_1 + x_3 \leq 1$, and $-x_1 + x_3 \leq 1$. For this set, there exist no functions G_α, H_β which satisfy the above requirements in a neighborhood of the point $(0,0,1)$.

If C is a cyrtohedron, then U may always be chosen so that (b') for all $x \in U$, (b) holds with x in place of $\bar{x}$; (c') if $A_0 \subset A_1 \subset A$ and $s(A_0) = s(A_1)$ then $Z(A_0) = Z(A_1)$; (d) for all $A_0 \subset A$, $Z(A_0)$ is a connected $(n-s(A_0))$-dimensional submanifold; and when this is done, we will say that $(G_\alpha(\alpha \in A), H_\beta(\beta \in B), U)$, or more briefly (G_α, H_β, U), is a *local representation* (abbreviated l.r.) for C.

Let (G_α, H_β, U) be a l.r., $x \in C \cap U$. Letting $A_+(x) = \{\alpha \in A : G_\alpha(x) = 0\}$, we define

$$L_C(x) = \{\zeta \in R^n : \zeta \cdot \nabla G_\alpha(x) = 0 \ \forall\ \alpha \in A_+(x),\ \zeta \cdot \nabla H_\beta(x) = 0\ \forall\ \beta \in B\}$$

$$N_C(x) = \left\{ \sum_{\alpha \in A_+(x)} a_\alpha \nabla G_\alpha(x) + \sum_{\beta \in B} b_\beta \nabla H_\beta(x) : a \in R_+^{A_+(x)} \text{ and } b \in R^B \right\}$$

$N_C(x)$ is the *normal cone* to C at x, and $L_C(x)$ is the *linear approximation* to C at x; the latter is the tangent space at x to the "face" (definition below) of C containing x. The *dimension* of C is defined to be $\dim C = n - |B|$. It does not depend on x, and none of these definitions depends on the particular local representation chosen.

For $x,y \in C$, define an equivalence relation $\sim$ by specifying $x \sim y$ if, and only if, there is a sequence $x = x_0, x_1, \ldots, x_p = y$ in C such that for each pair (x_i, x_{i+1}) $(i = 0,\ldots,p - 1)$, there is a l.r. (G_α, H_β, U) such that $Z(A) \supset \{x_i, x_{i+1}\}$. The equivalence classes under this relation are the *faces* of C.

A few examples help to clarify the latter definition: (a) the faces of a polyhedral convex set are the relative interiors of its "faces" in the usual sense (that is, subsets which are the intersection with some supporting hyperplane); (b) a submanifold $C \subset R^n$ has only one face; (c) if C is the hemisphere $C = \{x = (x_1,\ldots,x_n) \in R^n : |x| \leq 1 \text{ and } x_n \geq 0\}$, then C has four faces, corresponding to the choices of equality or strict inequality in the definition of C:

$$F_1 = \{x : |x| < 1 \text{ and } x_n > 0\}$$

$$F_2 = \{x : |x| = 1 \text{ and } x_n > 0\}$$

$$F_3 = \{x : |x| < 1 \text{ and } x_n = 0\}$$

$$F_4 = \{x : |x| = 1 \text{ and } x_n = 0\}$$

To state the optimality conditions, we need some more definitions. Consider a specific problem

$$\begin{aligned} &\min \quad f(x) \text{ over all } x \in R^n \\ &\text{such that } g(x) \leqq 0,\ h(x) = 0, \text{ and } x \in C \end{aligned} \tag{S}$$

If x is feasible for (S), the *independence criterion* (IC) is satisfied for (S) at x if for any $a \in R^m$ and $b \in R^k$ with $a_i = 0$ for all $i \notin I_+$,

$$\sum_{i=1}^{m} a_i \nabla g_i(x) + \sum_{j=1}^{k} b_j \nabla h_j(x) \in L_C(x)^{\perp} \text{ implies } 0 = a = b \tag{IC}$$

It is trivially satisfied if $m = k = 0$. If $C = R^n$, IC says that the gradients of the active constraints at x are linearly independent. More generally, IC says that the projections of the gradients of g_i, $i \in I_+$ and h_j at x onto $L_C(x)$ form a linearly independent set.

A set $M \subset R^n$ is a k-dimensional C^s *submanifold* ($s \geqq 1$) if for each $x \in M$ there is an open set $U \subset R^k$ and a C^s diffeomorphism Φ mapping U onto a neighborhood of x in M. For any $x = \Phi(q) \in M$, M_x = range $d\Phi(q)$ is the *tangent space* to M at x. If $f : R^n \to R$, then "$f|M$" denotes the restriction of f to M. For any $x \in R^n$, "$\nabla f(x)$" denotes the ordinary gradient of f at x, while "$\nabla(f|M)(x)$" denotes the gradient of $f|M$ at x, the latter being a linear function on M_x. If $\nabla(f|M)(x) = 0$ (i.e., if $\nabla f(x)$ is perpendicular to M_x), then x is a *critical point* for f on M, and in this case the *Hessian* for $f|M$ at $x = \Phi(q)$ is the bilinear function on M_x defined by

$$(\nabla^2(f|M)(x))(\bar{u},\bar{v}) = (\nabla^2(f\circ\Phi)(q))(u,v)$$

where $\bar{u} = d\Phi(x)u$, $\bar{v} = d\Phi(x)v$, and $\nabla^2(f\circ\Phi)(q)$ is the ordinary Hessian of $f\circ\Phi$. If $\nabla^2(f\circ\Phi)(q)$ is nonsingular, then x is a *nondegenerate critical* point for f on M.

Suppose henceforth that f, g, *and* h *are of class* C^2 *on* R^n, *and that* $C \subset R^n$ *is a cyrtohedron of class* C^2. We extend the definition of the *strong second order conditions* to the problem (S) by declaring a point

$\bar{w} = (\bar{x},\bar{y},\bar{z})$ with $\bar{x} \in C$, $\bar{y} \in R^m_+$, and $\bar{z} \in R^k$ to satisfy the conditions whenever the following conditions (SSOC) hold:

1. $\bar{x}$ is feasible for (S).
2. $-\nabla_x L(\bar{w}) \in \text{relint } N_C(\bar{x})$.
3. $i \in I$, $\bar{y}_i > 0$ if, and only if, $g_i(\bar{x}) = 0$.
4. The independence criterion for (S) holds at $\bar{x}$.
5. If F is the face of C containing $\bar{x}$, then $(\nabla^2_x(L|F)(\bar{w}))(\zeta,\zeta) > 0$ for all $\zeta \in R^n$ satisfying $0 \neq \zeta \in L_C(\bar{x})$, and $\zeta\cdot\nabla g_i(\bar{x}) = \zeta\cdot\nabla h_j(\bar{x}) = 0$ for all $i \in I_+$, and all j.

As before, we say the family (S(p)) is *full* provided the map $p' \to \nabla_w L(w,p') \in R^r$ has full rank r at all $(w,p) \in R^r \times P$. We now have covered all the preliminaries needed to state the final result.

THEOREM 3 *Let* $C \subset R^n$ *be a* d*-dimensional cyrtohedron of class* C^s, P *open,* f *of class* C^2 *and* g *and* h *of class* C^s *on* $R^n \times P$ *with* $s > \max\{1,d-k\}$. *If* (S(p)) *is full, there is a subset* $P_0 \subset P$ *with* $P\backslash P_0$ *negligible such that for all* $\bar{p} \in P_0$: *if* $\bar{x} \in C$ *is a local minimizer for* $(S(\bar{p}))$ *there exists* $(\bar{y},\bar{z}) \in R^m_+ \times R^k$ *satisfying* SSOC.

Of course, this result can be slightly improved in the manner of Corollaries 1 and 2.

V. COMPARISON WITH THE CLASSICAL CONDITIONS

For problems of the form (Q) we have seen that under mild assumptions, the classical strong conditions (SC)

1. $\bar{x}$ is feasible for (Q).
2. $\nabla f(\bar{x}) + \sum \bar{y}_i \nabla g_i(\bar{x}) + \sum \bar{z}_j \nabla h_j(\bar{x}) = 0$.
3. Strict complementary slackness: $\bar{y}_i > 0 \iff g_i(\bar{x}) = 0$.
4. The gradients of the active constraints, i.e., $\{\nabla g_i(\bar{x}) : i \in I_+\} \cup \{\nabla h_j(\bar{x}) : j = 1,\ldots,k\}$, form a linearly independent set.
5. For any $\zeta \in R^n$ satisfying $\zeta \neq 0$, $\zeta\cdot\nabla g_i(\bar{x}) = 0 \ \forall\, i \in I_+$, and $\zeta\cdot\nabla h_j(\bar{x}) = 0$, $j = 1,\ldots,k$, we have $\zeta'[\nabla^2 f(\bar{x}) + \sum \bar{y}_i \nabla^2 g_i(\bar{x}) + \sum \bar{z}_j \nabla^2 h_j(\bar{x})]\zeta > 0$

are generically necessary for optimality in families of problems containing (Q) (cf. Theorems 1 and 2), and that for problems of the form (S) (i.e., families with fixed cyrtohedron constraints), the more general conditions SSOC are generically necessary for optimality.

Locally, the fixed set C can be represented by inequality and equality constraints; if (G_α, H_β, U) is a local representation for C, then $C \cap U = \{x \in U : G_\alpha(x) \leqq 0, \alpha \in A, H_\beta(x) = 0, \beta \in B\}$. So, at least locally, (S) is equivalent to a problem (Q') of the type (Q) (i.e., without "fixed" constraints)

$$\begin{aligned} &\min \quad f(x) \\ &\text{s.t.} \quad g_i(x) \leqq 0,\ i = 1,\ldots,m,\ h_j(x) = 0,\ j = 1,\ldots,k, \\ &G_\alpha(x) \leqq 0,\ \alpha \in A,\ H_\beta(x) = 0,\ \beta \in B \end{aligned} \tag{Q'}$$

It is natural to ask what the relationship is between the conditions SSOC for (S) and SC for (Q').

In most cases, the two sets of conditions are essentially equivalent in the following sense. If $(\bar{x},\bar{y},\bar{a},\bar{z},\bar{b}) \in R^n \times R^m_+ \times R^A_+ \times R^k \times R^B$ satisfies SC for (Q'), then $(\bar{x},\bar{y},\bar{z})$ satisfies SSOC for (S). If $(\bar{x},\bar{y},\bar{z}) \in R^n \times R^m_+ \times R^k$ satisfies SSOC for (S), then it is possible to find $\bar{a} \in R^A_+$ and $\bar{b} \in R^B$ such that $(\bar{x},\bar{y},\bar{a},\bar{z},\bar{b})$ satisfies SC1, 2, 3, and for any such $\bar{a}$ and $\bar{b}$, SC5 will automatically hold for (Q'). *However*, SC4 *may fail*. For example, if C is a four-sided pyramid in R^3 with apex $\bar{x}$, SC4 can never be satisfied for (Q') because no set of four vectors in R^3 can be linearly independent. However, SSOC4 can (and usually will) be satisfied at $\bar{x}$. In fact, $(\bar{x},\bar{y},\bar{z})$ will satisfy SSOC4 if and only if the projections onto $L_C(\bar{x})$ of the gradients of the (nonfixed) constraints active at $\bar{x}$ are linearly independent. But $L_C(\bar{x}) = \{0\}$ in this case, so SSOC4 merely says that *there are no active constraints* at $\bar{x}$. Of course, one would expect the generic conditions to assert this. If $k > 0$, one would expect the apex of the pyramid to be a minimizer with probability zero. If $k = 0$, it is not unusual that the apex should be a minimizer, but one would expect one or more of the inequality constraints to be active there only with probability zero.

In the most common cases, such as $C = R^n_+$, the set C will be expressible as the set of points which satisfy a finite number of equality and inequality constraints with linearly independent gradients (cf. section III, example (d) under "examples of cyrtohedra"). Then, the two sets of conditions are essentially the same. The main difference is that in the SSOC formulation, no multipliers are associated with the constraints defining the cyrtohedron.

We also remark that the SSOC formulation suggests what the generic conditions should look like if we generalize them to a wider class of fixed sets C. Consider, for example, the set

$$C = \{x = (x_1, x_2, x_3) \in R^3 : |x| \leqq 1 \text{ and } x_1 + x_2 + x_3 \geqq |x|\}$$

Because no local representation exists for C near $\bar{x} = 0$, C is not a cyrtohedron. But, like a cyrtohedron, C can be partitioned into "faces" (four in this case) that are submanifolds, and $N_C(x)$ and $L_C(x)$ have obvious meanings, so the conditions SSOC, as stated above, are still meaningful. In fact, C has all the properties that are required for our proof of the genericity of SSOC. For such a set C, it would be impossible to reformulate the problem (S) as a problem in the form of (Q'), so the old conditions SC have no bearing here, although the new conditions SSOC would apply and can be shown to be generically necessary for optimality. We do not know if there is a "natural" broader class to which our results apply. The above example suggests conditions should be generic for sets C that look (in some sense) locally like the intersection of a cone with a neighborhood of the origin. One possible class would be those sets C such that each $x \in C$ has a neighborhood U such that for some diffeomorphism ϕ, and some closed convex cone K, $\phi(x) = 0$ and $\phi(C \cap U) = \phi(U) \cap K$. For this class, the proof of the genericity of the above conditions does indeed go through. However, this is not as broad a class as we would like; it does not seem even to include the class of cyrtohedra.

REFERENCES

1. A. V. Fiacco, Sensitivity analysis for nonlinear programming using penalty methods, *Math. Programming 10*(1976), 287-311.

2. A. V. Fiacco and G. P. McCormick, *Nonlinear Programming: Sequential Unconstrained Minimization Techniques*, New York: Wiley, 1968.

3. M. R. Hestenes, *Optimization Theory, the Finite Dimensional Case*, New York: Wiley, 1975.

4. M. J. D. Powell, A method for nonlinear constraints in minimization problems, in *Optimization* (R. Fletcher, ed.), New York: Academic Press, 1972.

5. S. M. Robinson, Perturbed Kuhn-Tucker points and rates of convergence for a class of nonlinear programming algorithms, *Math. Programming 7* (1974), 1-16.

6. R. T. Rockafellar, Augmented Lagrange multiplier functions and duality in nonconvex programming, *SIAM J. Control 12*(1974), 268-285.

7. J. E. Spingarn, Fixed and variable constraints in sensitivity analysis, *SIAM J. Control 18*(1980).

8. J. E. Spingarn, On optimality conditions for structured families of

nonlinear programming problems, forthcoming.

9. J. E. Spingarn, Generic conditions for optimality in constrained minimization problems, Dissertation, Department of Mathematics, University of Washington, 1977.

10. J. E. Spingarn and R. T. Rockafellar, The generic nature of optimality conditions in nonlinear programming, *Math. of O.R.* 4(1979).

Chapter 2 ON THE CONVERGENCE OF SEQUENCES OF CLOSED FUNCTIONS*

ROGER J. B. WETS / University of Kentucky, Lexington, Kentucky

ABSTRACT

A sequence of closed (lower semicontinuous and proper) functions is said to e-converge if their epigraphs converge. We study the relations between e-convergence and (pointwise) p-convergence. The main result shows that when the collection is equi-lower semicontinuous, e- and p-convergence imply each other. The implications for optimization problems are sketched out.

INTRODUCTION

To solve difficult (or infinite dimensional) problems, one often needs to replace the original problem by an approximate one. The fact that every approximate does not provide an approximating optimal solution is well documented. In this chapter, we characterize a class of approximate problems that are guaranteed to generate "approximate" optimal solutions. The results recorded here generalize those derived earlier for *convex* optimi-

---

*Supported in part by the National Science Foundation under Grant ENG-79003731.

zation problems. First, we recall some general facts about the convergence of sequences of closed sets in R^n. We then introduce and characterize the notion of e-convergence (i.e., convergence of the epigraphs) for sequences of closed functions with values in the extended reals. Next, we study the relations between e-convergence and (pointwise) p-convergence. Finally, the connections between e-convergence and the convergence of optimal solutions of optimization problems is made explicit in the last section.

I. SEQUENCES OF CLOSED SETS

A sequence of closed subsets $\{F_\nu, \nu \in N\}$ of R^n is said to converge to a closed set F if

$$ls\ F_\nu = F = li\ F_\nu \tag{1}$$

in which case, we write $\lim F_\nu = F$, where

$$li\ F_\nu = \{x = \lim x_\nu \mid x_\nu \in F_\nu, \nu \in N\}$$

and

$$ls\ F_\nu = \{x = \lim x_\mu \mid x_\mu \in F_\mu, \mu \in M \subset N\}$$

By M, we always denote an infinite ordered subset of the countable index set N. This will allow us to distinguish between statements that need to hold for all--but a finite number of--elements in the sequence and those involving only some countable number of elements in the sequence. The sets $li\ F_\nu$ and $ls\ F_\nu$ are closed and are called the *liminf* and *limsup* of the sequence $\{F_\nu, \nu \in N\}$, respectively. The next theorem, which provides a number of criteria for the convergence of sequences of closed sets, is proved in [1] and [2]. By d we denote the *metric* on R^n; $B_r(c) = \{y \mid d(c,y) \leq r\}$ is the *closed ball* of radius r and center c, and $B_r^\circ(c) = \{y \mid d(c,y) < r\}$ is the corresponding open ball. Let D be a subset of R^n, then $d(x,D)$ is the distance from x to D, i.e., $d(x,D) = \inf[d(x,y) \mid y \in D]$. Note that $d(x,D)$ is finite unless $D = \emptyset$, in which case $d(x,\emptyset) = +\infty$. By $\varepsilon D = \{y \mid d(y,D) < \varepsilon\}$ we denote the open ε-*neighborhood* of D--with $\varepsilon\emptyset = R^n \setminus B_{\varepsilon^{-1}}(0)$--and by $D^{r,x} = D \cap B_r(x)$ the intersection of D with a ball of radius r and center x. We write simply D^r for $D^{r,0} = D \cap B_r(0)$. The *Hausdorff distance* $h(C,D)$ between the sets C and D is defined as follows:

$$h(C,D) = \begin{cases} 0 & \text{if } C = D = \emptyset \\ +\infty & \text{if } C = \emptyset \text{ but not } D \text{ or vice versa} \\ \sup[\sup_{x\in C} d(x,D),\ \sup_{y\in D} d(y,C)] \end{cases}$$

The ρ-*distance* between the sets C and D is given by

$$h_\rho(C,D) = h(C^\rho, D^\rho)$$

THEOREM 1 *Suppose that* $\{F ; F_\nu, \nu \in N\}$ *is a collection of closed subsets of* R^n. *Then the following are equivalent:*

(i) $\lim_\nu F_\nu = F$;

(ii) K *compact and* $F \cap K = \emptyset$ *implies that* $F_\nu \cap K = \emptyset$ *for* ν *sufficiently large, and* F *open and* $F \cap G \neq \emptyset$ *implies that* $F_\nu \cap G \neq \emptyset$ *for* ν *sufficiently large;*

(iii) $F \cap B_r(c) = \emptyset$ *implies that* $F_\nu \cap B_r(c) = \emptyset$ *for* ν *sufficiently large, and* $F \cap B_r^\circ(c) \neq \emptyset$ *implies that* $F_\nu \cap B_r^\circ(c) \neq \emptyset$ *for* ν *sufficiently large;*

(iv) $\lim d(x,F_\nu) = d(x,F)$ *for all* $x \in R^n$;

(v) $\lim(F \setminus \varepsilon F_\nu) = \lim(F_\nu \setminus \varepsilon F) = \emptyset$ *for all* $\varepsilon > 0$;

(vi) *for all* $x \in R^n$, *and* $\varepsilon > 0$, $r > 0$, $F^{r,x} \subset \varepsilon F_\nu$ *and* $F_\nu^{r,x} \subset \varepsilon F$ *for* ν *sufficiently large;*

(vii) *for all* $x \in R^n$, $\lim_{r\uparrow\infty} li\ F_\nu^{r,x} \subset F \subset \lim_{r\uparrow\infty} ls\ F_\nu^{r,x}$.

If the sets $\{F ; F_\nu, \nu \in N\}$ *are also convex, then the preceding conditions are equivalent to*

(viii) *there exists* $\rho' > 0$ *such that* $\lim h_\rho(F_\nu,F) = 0$ *for all* $\rho \geqq \rho'$.

Finally, if the sets $\{F ; F_\nu, \nu \in N\}$ *are compact (convex or not) then* $F = \lim F_\nu$ *if and only if*

(ix) $\lim h(F_\nu,F) = 0$

II. e-CONVERGENCE

We consider collections $\{f ; f_\nu, \nu \in N\}$ of *closed functions* with domain R^n and range $R \cup \{+\infty\}$. A function is closed if it is lower semicontinuous and finite for at least some x in R^n, i.e., not identically $+\infty$. By D (D_ν resp.) we denote the *effective domain* of the function f (f_ν resp.), i.e.,

$$D = \{x \in R^n \mid f(x) < +\infty\}$$

Its *epigraph* E (E_ν resp.) is given by

$$E = \{(x,\alpha) \in R^{n+1} \mid f(x) \leqq \alpha\}$$

A function f is closed if and only if E is a nonempty closed subset of R^{n+1}; this also implies that D is nonempty but not necessarily closed.

Two different types of convergence for sequences of closed functions play a key role in the study of the convergence of the solutions to optimization problems. We say that the f_ν *pointwise converge* to f, and we write $f_\nu \underset{p}{\to} f$, if for all x in R^n, $\lim_\nu f_\nu(x) = f(x)$ where we allow $+\infty$ as limit value as well as element value in the sequence $\{f_\nu(x), \nu \in N\}$. The sequence of functions f_ν is said to *e-converge* to the function f if their epigraphs converge to the epigraph of f, i.e., $\lim E_\nu = E$.

PROPOSITION 1 *Suppose that* $\{f ; f_\nu, \nu \in N\}$ *are closed functions. Then* $f_\nu \underset{e}{\to} f$ *if and only if (i) for all* $x \in D$, $x = \lim x_\mu$ *implies that* $f(x) \leqq \liminf f_\mu(x_\mu)$; *(ii) to each* $x \in R^n$ *one can associate a sequence* $\{x_\nu, \nu \in N\}$ *converging to* x *such that* $\limsup f_\nu(x_\nu) \leqq f(x)$.

PROOF. We show that (i) or (ii) implies and is implied by $ls\ E_\nu \subset E$ ($E \subset li\ E_\nu$ resp.).

Let $x = \lim x_\mu$, $\alpha_\mu = f(x_\mu)$ and $\alpha = \liminf \alpha_\mu$. The elements converging to α determining a subsequence such that $(x,\alpha) = \lim[(x_\mu,\alpha_\mu), \mu \in M']$. If $f_\nu \underset{e}{\to} f$ then $E \supset ls\ E_\nu$ and since the (x_ν,α_ν) belong to E_μ for all $\mu \in M'$, it follows that $(x,\alpha) \in E$, i.e., $\alpha \geqq f(x)$. On the other hand, if $\{(x_\mu,\alpha_\mu) \in E_\mu, \mu \in M\}$ is a sequence of points converging to (x,α), from (i) it follows that $(x,\alpha) \in E$ since

$$\alpha = \lim \alpha_\mu \geqq \liminf f_\mu(x_\mu) \geqq f(x)$$

If $f_\nu \underset{e}{\to} f$ then $E \subset li\ E_\nu$ and thus to each $x \in D$ we can associate a sequence $\{(x_\nu,\alpha_\nu) \in E_\nu, \nu \in N\}$ such that $\lim(x_\nu,\alpha_\nu) = (x,f(x))$. Since for each ν, $\alpha_\nu \geqq f_\nu(x_\nu)$ it then also follows that $\limsup f_\nu(x_\nu) \leqq f(x)$. This implies (ii) since the inequality is trivially satisfied if $x \notin D$, i.e., when $f(x) = +\infty$. On the other hand, if $(x,\alpha) \in E$ then (ii) yields the existence of a sequence such that $f(x) \geqq \limsup f_\nu(x_\nu)$ which in turn implies that for all $\varepsilon > 0$ and ν sufficiently large

$$\alpha \geqq f(x) \geqq f_\nu(x_\nu) - \varepsilon$$

Let (x_ν,α_ν) be the projection of (x_ν,α) on $\{(x_\nu,\alpha) \mid \alpha \geqq f(x_\nu)\}$. From the above, it follows that given $\varepsilon > 0$ there exists ν_ε such that $d(\alpha,\alpha_\nu) \leqq \varepsilon$ for all $\nu \geqq \nu_\varepsilon$. Hence $(x,\alpha) = \lim(x_\nu,\alpha_\nu)$ and consequently *li* $E_\nu \supset E$.

III. RELATIONS BETWEEN e-CONVERGENCE AND p-CONVERGENCE

Neither type of convergence implies the other, not even when the functions are convex; in fact, not even when they are linear on their effective domain and the effective domain is defined by a finite number of linear relations. To see this, consider for example the sequence of closed convex ("linear") functions defined as follows: for $\nu = 1,2,\ldots$

$$g_\nu(x_1,x_2) = \begin{cases} \nu x_2 & \text{if } x_1 \geqq 0 \text{ and } x_1 + \nu x_2 \geqq 0 \\ +\infty & \text{otherwise} \end{cases}$$

Then $g_\nu \xrightarrow[p]{} g$ but $g_\nu \xrightarrow[e]{} g'$ where

$$g(x_1,x_2) = \begin{cases} 0 & \text{if } x_1 \geqq 0,\ x_2 = 0 \\ +\infty & \text{otherwise} \end{cases}$$

and

$$g'(x_1,x_2) = \begin{cases} -x_1 & \text{if } x_1 \geqq 0,\ x_2 = 0 \\ +\infty & \text{otherwise} \end{cases}$$

Equivalence between p- and e-convergence demands an equi-continuity condition. Such a condition was first introduced by G. Salinetti and the author in [3] in our study of the convergence of closed *convex* functions. Here it is shown that the basic results can be extended to the nonconvex case provided that D is closed, i.e., D = cl D.

DEFINITION. A collection of closed functions $\{f\,;\, f_\nu,\ \nu \in N\}$ is said to be *equi-lower semicontinuous* (*equi-l.s.c.*) if the following conditions are satisfied:

(3α) to each $x \in D$ and $\varepsilon > 0$, there corresponds V a neighborhood of x such that for all ν in N, $f_\nu(x) - \varepsilon \leqq f_\nu(y)$ for all $y \in V$;

(3β) to each $x \in D$, there corresponds ν_x such that $x \in D_\nu$ for all $\nu \geqq \nu_x$;

(3γ) $\{f_\nu,\ \nu \in N\}$ converge uniformly to $+\infty$ on every compact subset of $R^n \setminus \text{cl } D$, the complement of the closure of D.

It is easy to verify that both collections $\{g\,;\, g_\nu,\ \nu \in N\}$ and $\{g'\,;\, g_\nu,\ \nu \in N\}$ in the example (preceding the definition) fail to satisfy condition (3α). If $f_\nu \underset{p}{\to} f$ then (3β) is automatically satisfied but does not necessarily hold if $f_\nu \underset{e}{\to} f$. Consider $h_\nu(x) = \infty$ unless $x \in [\nu^{-1},1]$ in which case $h_\nu(x) = 0$. Then $h_\nu \underset{e}{\to} h$ where $h(x) = \infty$ unless $x \in [0,1]$ in which case $h(x) = 0$; but $\lim h_\nu(0) = +\infty \neq 0 = h(0)$. On the other hand, if $f_\nu \underset{e}{\to} f$ then (3γ) automatically follows; however, not necessarily if the f_ν converge pointwise to f. For example, let $j_\nu(x_1,x_2) = \infty$ unless $\nu x_2 = x_1$ and then $j_\nu(x_1,x_2) = 0$. Then $j_\nu \underset{p}{\to} j$ where $j \equiv \infty$ on $R^2 \setminus \{(0,0)\}$ and $j(0,0) = 0$. On the segment $[(0,1,),(1,0)]$ condition (3γ) fails and in fact, the j_ν do not e-converge to j but to j', where $j'(0,x_2) = 0$ for any value of x_2 and $j'(x_1,x_2) = +\infty$ if $x_1 \neq 0$.

We start by deriving some of the implications of e- and p-convergence.

LEMMA 1 *Suppose that* $\{f\ ;\ f_\nu,\ \nu \in N\}$ *are closed functions and* $f_\nu \underset{p}{\to} f$. *Then* (i) $E \subset$ *li* E_ν; (ii) *the collection* $\{f\ ;\ f_\nu,\ \nu \in N\}$ *satisfies* (3β).

PROOF. (ii) follows directly from the definition of pointwise convergence. To obtain (i) simply note that every $(x,\alpha) \in E$ can be obtained as the limit of the sequence $\{(x,\alpha_\nu),\ \nu \in N \mid \alpha_\nu = \max(\alpha, f_\nu(x))\}$ and clearly $(x,\alpha_\nu) \in E_\nu$.

LEMMA 2 *Suppose that* $\{f\ ;\ f_\nu,\ \nu \in N\}$ *are closed functions and* $f_\nu \underset{e}{\to} f$. *Then* (i) *for all* $x \in D$, $\liminf f_\nu(x) \geq f(x)$; (ii) *the collection* $\{f\ ;\ f_\nu,\ \nu \in N\}$ *satisfies* (3γ).

PROOF. (i) follows directly from Proposition 1 (i) with the sequence $\{x_\nu = x,\ \nu \in N\}$. To prove (ii) we argue by contradiction. Suppose K is a compact subset in the complement of cl D on which the f_ν fail to converge uniformly to $+\infty$. This means that given any number η arbitrarily large, there is an infinite number of indices, say all $\mu \in M$, such that $f_\mu(x_\mu) \leq \eta$ for some $x_\mu \in K$. The compactness of K in turn implies that the $\{x_\mu\}$ admit a cluster point $x \in K$ such that $x = \lim(x_\mu,\ \mu \in M')$. It then follows that $x \in D$, since $\liminf(f_\mu(x_\mu),\ \mu \in M') \leq \eta$ and *ls* $E_\nu \subset E$ implies that $(x, \liminf f_\nu(x_\nu)) \in E$. Hence $K \cap D \neq \emptyset$.

LEMMA 3 *Suppose that* $\{f\ ;\ f_\nu,\ \nu \in N\}$ *is a collection of closed functions satisfying* (3γ) *and such that* $f_\nu \underset{p}{\to} f$. *Then*

$$D \subset \text{prj } ls\ E_\nu \subset \text{cl } D$$

where prj *ls* $E_\nu = \{x \mid (x,\alpha) \in ls\ E_\nu\}$.

PROOF. The first inclusion follows from Lemma 1 (i) and the fact that $li\ E_\nu \subset ls\ E_\nu$. To establish the second inclusion, we show that if $x \notin \text{cl}\ D$ then $x \notin \text{prj}\ ls\ E_\nu$. Suppose not, i.e., there exists $(x,\alpha) \in ls\ E_\nu$ with $x \notin \text{cl}\ D$. Condition (3γ) implies that if K is a compact neighborhood of x contained in $R^n \setminus \text{cl}\ D$, then given any real number η (arbitrarily large) we have that for ν sufficiently large

$$(V \times [-\infty,\eta]) \cap E_\nu = \emptyset$$

In particular, this holds for $\eta > \alpha$ and consequently (x,α) could not belong to $ls\ E_\nu$ or equivalently $x \notin \text{prj}\ ls\ E_\nu$.

THEOREM 2 *Suppose that* $\{f\ ;\ f_\nu,\ \nu \in N\}$ *are closed functions with* $D = \text{cl}\, D$. *Then any two of the following statements imply the other: (i) the collection* $\{f\ ;\ f_\nu,\ \nu \in N\}$ *is equi-l.s.c.; (ii)* $f_\nu \underset{p}{\to} f$*; (iii)* $f_\nu \underset{e}{\to} f$. *The condition on* D*, the effective domain of* f*, can be dropped if the functions are also convex.*

PROOF. (ii) + (iii) ⟹ (i). In view of Lemmas 1 (ii) and 2 (ii) it suffices to show that condition (3α) is satisfied. To the contrary suppose that (3α) fails for some $x \in D$; this means that there exists $\varepsilon > 0$ such that in every neighborhood V there exists y such that for some ν, $f_\nu(x) > f_\nu(y) + \varepsilon$. Take $\{V_k,\ k \in N\}$ a sequence of nested neighborhoods of x such that $\cap_k V_k = \{x\}$, and let ν_k be the index for which the required inequality fails at $y_k \in V_k$. From p-convergence, it follows that

$$f(x) \geqq \liminf f_{\nu_k}(y_k) + \varepsilon$$

On the other hand, the $x = \lim y_k$ and thus from e-convergence, via Proposition 1 (i), we know that

$$f(x) \leqq \liminf f_{\nu_k}(y_k)$$

which contradicts the preceding inequality.

(i) + (iii) ⟹ (ii). In view of Lemma 2 (i), it suffices to show that $\limsup f_\nu(x) \leqq f(x)$ when $x \in D$ since otherwise the inequality is clearly satisfied. Proposition 1 (ii) guarantees the existence of a sequence $\{x_\nu,\ \nu \in N\}$ such that $x = \lim x_\nu$ with $\limsup f_\nu(x_\nu) \leqq f(x)$. For ν sufficiently large, it follows from equi-l.s.c., more precisely from (3β) that $x \in D_\nu$ and from (3α) that $f_\nu(x) - \varepsilon \leqq f_\nu(x_\nu)$. Thus, for all $\varepsilon > 0$, we have that

$$\limsup f_\nu(x) - \varepsilon \leqq \limsup f_\nu(x_\nu) \leqq f(x)$$

which yields the desired inequality.

(i) + (ii) $\Rightarrow$ (iii). In view of Lemma 1 (i), it suffices to show that $ls\ E_\nu \subset E$. Choose any $(x,\alpha) \in ls\ E_\nu$; then $(x,\alpha) = \lim[(x_\mu,\alpha_\mu), \mu \in M \mid (x_\mu,\alpha_\mu) \in E_\mu]$. Lemma 3 and D = cl D imply that $x \in D$. From the equi-l.s.c., more precisely (3α), it then follows that given any $\varepsilon > 0$, for μ sufficiently large $f(x) - \varepsilon \leqq f_\mu(x_\mu) \leqq \alpha_\mu$ and thus

$$f(x) = \liminf f(x) \leqq \liminf f_\mu(x_\mu) \leqq \lim \alpha_\mu = \alpha$$

which implies that $(x,\alpha) \in E$.

It remains to show that in the convex case $D = \text{prj}\ ls\ E_\nu$ when (i) and (ii) are satisfied. Suppose not, then in view of Lemma 3, there exists $x \in \text{cl}\ D \setminus D$ and $\alpha \in R$ such that $(x,\alpha) \in ls\ E_\nu$. Note that $C = \text{con}\{(x,\alpha), E\}$ is contained in $ls\ E_\nu$ since it is convex and $E \subset li\ E_\nu \subset ls\ E_\nu$ since $f_\nu \underset{p}{\to} f$, cf. Lemma 1 (i). Choose $(x',\alpha') \in C \subset ls\ E_\nu$ such that $x' \in D$ and $\alpha' < f(x')$; such a point exists because $(x,\alpha) \notin E$ and E is closed and convex. Also, there exists $\{(x'_\mu,\alpha'_\mu) \in E_\mu, \mu \in M\}$ such that $(x',\alpha') = \lim(x'_\mu,\alpha'_\mu)$. Now by equi-l.s.c., in particular (3β) and (3α), for any given $\varepsilon > 0$ and μ sufficiently large, we have that

$$\alpha' - \varepsilon < f(x') - \varepsilon \leqq f_\mu(x'_\mu) \leqq \alpha'_\mu$$

and consequently

$$\alpha' < f(x') \leqq \lim \alpha'_\mu = \alpha'$$

an evident contradiction.

The condition cl D = D cannot be ignored in the nonconvex case. The following example illustrates the difficulties one may encounter. For $\nu = 1,2,\ldots,$ let

$$f_\nu(x) = \begin{cases} x^{-1} & \text{if } |x| \neq \nu^{-1} \\ 0 & \text{if } |x| = \nu^{-1} \end{cases}$$

Then clearly $f_\nu \underset{p}{\to} f$ with $f(x) = x^{-1}$; it being understood that $f(0) = 0^{-1} = +\infty$. Note that the collection $\{f\ ;\ f_\nu, \nu \in N\}$ is equi-l.s.c., but the f_ν do not e-converge to f since $f_\nu \underset{e}{\to} g$, with

$$g(x) = \begin{cases} x^{-1} & \text{if } x \neq 0 \\ 0 & \text{if } x = 0 \end{cases}$$

as can easily be verified. Note that the collection $\{g ; f_\nu, \nu \in N\}$ is not equi-l.s.c.

IV. APPLICATIONS TO OPTIMIZATION PROBLEMS

Again $\{f ; f_\nu, \nu \in N\}$ will represent a family of (closed) functions. By A_ν (A resp.) we denote the argmin f_ν (argmin f resp.), i.e., those elements of R^n achieving the minimum of f_ν (f resp.). If the infimum of f_ν (f resp.) is not attained then A_ν (A resp.) is empty. Let $m_\nu = \inf f_\nu$ ($m = \inf f$ resp.); we set m_ν (m resp.) equal to $-\infty$ if the infimum is not bounded below. The two next theorems provide the main justification for the introduction of the concept of e-convergence in the study of the convergence properties of the optimal solutions of optimization problems.

THEOREM 3 *Suppose that* $\{f ; f_\nu, \nu \in N\}$ *are closed functions such that* $f_\nu \underset{e}{\rightarrow} f$. *Then* $\text{ls } A_\nu \subset A$.

PROOF. The inclusion is trivially true if the A_ν are empty for all ν (except possibly for a finite number). Let us thus assume, without loss of generality, that the A_ν are nonempty for all $\nu \in N$ and that there is a sequence $\{x_\mu \in A_\mu, \mu \in M\}$ converging to x. We need to show that $x \in A$. We consider the cases $m = \inf f > -\infty$ and $m = -\infty$ separately.

When f is not bounded below, it means that given any real number η, we can find y such that $(y,\eta) \in E$. Since the $f_\nu \underset{e}{\rightarrow} f$, it follows that there exists a sequence $\{(y_\nu,\eta_\nu) \in E_\nu, \nu \in N\}$ converging to (y,η), in particular we have that $(y,\eta) = \lim(y_\mu,\eta_\mu)$. Since $x_\mu \in A_\mu$, it follows that

$$m_\mu = f_\mu(x_\mu) \leq \eta_\mu \qquad \text{for all } \mu \in M$$

and via Proposition 1 (i), we then get

$$f(x) \leq \liminf f_\mu(x_\mu) \leq \lim \eta_\mu = \eta$$

Since the above holds for every η it implies that $f(x) = -\infty$ contradicting the assumption that f is closed.

When $m = \inf f$ is finite but $x \notin A$--possibly because A is empty--it means that to any given $\varepsilon > 0$ sufficiently small, there corresponds y such that

$$\eta = f(y) < m + \varepsilon < f(x)$$

Since $(y,\eta) \in E$ and $\lim E_\nu = E$, there is a sequence of points $\{(y_\nu,\eta_\nu) \in E_\nu,\ \nu \in N\}$ converging to (y,η). Again, $x_\mu \in A_\mu$ and thus for $\mu \in M$,

$$m_\mu = f_\mu(x_\mu) \leqq \eta_\mu$$

and consequently, in view of Proposition 1 (i), we have that

$$f(x) \leqq \liminf f_\mu(x_\mu) \leqq \lim \eta_\mu = \eta < f(x)$$

an evident contradiction.

Note that Theorem 3 only implies the existence of a minimum of f if also *ls* A_ν is nonempty. It follows from Theorem 1 (v) and the definition of an ε-neighborhood of $\emptyset$ that *ls* $A_\nu = \emptyset$ only if for all $r > 0$, *ls* $A_\nu^r = \emptyset$. For each r, the sets A_ν^r are uniformly bounded and thus *ls* A_ν^r is empty if and only if the A_ν^r are all empty for ν sufficiently large. From these observations we obtain immediately the following:

COROLLARY 1 *Suppose that* $\{f\ ;\ f_\nu,\ \nu \in N\}$ *are closed functions such that* $f_\nu \underset{e}{\to} f$. *Then* A *is nonempty whenever there exists* $r > 0$ *and* $M \subseteq N$ *such that* A_μ^r *is nonempty for all* $\mu \in M$.

COROLLARY 2 *Suppose that* $\{f\ ;\ f_\nu,\ \nu \in N\}$ *are closed functions such that* $f_\nu \underset{e}{\to} f$ *and ls* A_ν *is nonempty. Then* $\lim m_\nu = m = \text{Min } f$.

COROLLARY 3 *Suppose that* $\{f\ ;\ f_\nu,\ \nu \in N\}$ *are closed functions such that* $f_\nu \underset{e}{\to} f$, *that the minima of the* f_ν *and of* f *are attained at* x_ν *and* x, *respectively, and are unique. Suppose moreover that the* x_ν *are bounded. Then* $m = \lim m_\nu$ *and* $x = \lim x_\nu$.

Clearly, the infima m_ν of the functions f_ν might converge to the infima m of the function f even when the A_ν and/or A are empty. One such result is derived here below. Let g be an arbitrary function defined on R^n and with values in $R \cup \{+\infty\}$ and denote by clcon g the function whose epigraph is the closure of the convex hull of the epigraph of g. Then inf g = inf (clcon g). This is evident if inf g = $-\infty$, otherwise simply note that every point in the epigraph of clcon g must have been obtained as the limit of a sequence of points which are convex combinations of elements in the epigraph of g and thus the hyperplane $\{(x,\alpha) \mid x \in R^n,\ \alpha = \inf g\}$ bounds the epigraph of clcon g from below. The function clcon g inherits many of the properties of g, for example if g is inf-compact--i.e., the function g is closed, and for all $\alpha \in R$ the level sets $\{x \mid f(x) \leq \alpha\}$ are compact--then clcon g is inf-compact; if the effective domain of g has nonempty interior then so does the effective domain of g; finally, note that if g attains its minimum then so does clcon g, in fact we always have that:

$$\text{argmin } g \subset \text{argmin } (\text{clcon } g)$$

The (convex) conjugate of a function g is by definition

$$g^*(v) = \sup\{x \cdot v - g(x)\}$$

PROPOSITION 2 *Suppose that* $\{f\ ;\ f_\nu,\ \nu \in N\}$ *is a collection of inf-compact convex functions such that* $f_\nu \underset{e}{\to} f$. *Then* $\lim m_\nu = m$.

PROOF. Inf-compactness implies not only that the infima are attained but also that f^* and f_ν^* are continuous at 0. Moreover, a basic result in the convergence theory for convex functions shows that conjugation is bicontinuous with respect to the e-convergence topology, see [4] and [5]. Thus, $f_\nu^* \underset{e}{\to} f^*$ which together with the continuity of the f_ν^* and of f^* at 0 implies that $\lim m_\nu = \lim -f_\nu^*(0) = -f^*(0) = m$.

REFERENCES

1. G. Salinetti and R. Wets, On the convergence of sequences of closed sets, *Topology Proceedings*, *4*(1979).
2. G. Salinetti and R. Wets, On the convergence of sequences of convex sets in finite dimensions, *SIAM Review*, *21*(1979), 18-33.
3. G. Salinetti and R. Wets, On the relation between two types of convergence for convex functions, *J. Math. Anal. Appl.*, *60*(1977), 211-226.
4. R. Wijsman, Convergence of sequences of convex sets, cones and functions. II, *Trans. Amer. Math. Soc.*, *123*(1966), 32-45.
5. D. Walkup and R. Wets, Continuity of some convex-cone valued mappings, *Proc. Amer. Math. Soc.*, *18*(1967), 229-235.

Chapter 3 OPTIMAL VALUE DIFFERENTIAL STABILITY RESULTS FOR GENERAL INEQUALITY CONSTRAINED DIFFERENTIABLE MATHEMATICAL PROGRAMS

ANTHONY V. FIACCO* / The George Washington University, Washington, D.C.
WILLIAM P. HUTZLER / The Rand Corporation, Washington, D.C.

I. INTRODUCTION

This chapter summarizes the key results that were presented and validated by the authors in [11]. We first provide a brief perspective.

Using point-to-set maps, Berge [5] derived conditions sufficient for the semicontinuity of the optimal value function for programs with constraint set perturbations, and provided a general framework for some of the earliest work on the variation of the "perturbation function," i.e., the optimal objective function value, with changes in a parameter appearing in the right-hand side of the constraints.

Evans and Gould [8] gave conditions guaranteeing the continuity of the perturbation function when the constraints are functional inequalities. Greenberg and Pierskalla [16] extended the work of Evans and Gould to obtain results for general constraint perturbations and obtained some initial results for programs with equality constraints. In [18], Hogan established conditions sufficient for the continuity of the perturbation function of a convex program, and in [19] gave conditions implying the

---

*Dr. Fiacco's research was supported by the U.S. Army Research Office.

continuity of the optimal value function of a nonconvex program when a parameter appears in the objective function.

The first- and second-order variation of the optimal value of a general nonlinear program under quite arbitrary parametric perturbations has been investigated by Hogan [18], Armacost and Fiacco [1,2,3], Fiacco [10], and Fiacco and McCormick [12]. In [2] the optimal value function is shown, under strong conditions, to be twice continuously differentiable, with respect to the problem parameters, with its parameter gradient (Hessian) equal to the gradient (Hessian) of the Lagrangian of the problem. Armacost and Fiacco [1] have also obtained first- and second-order expressions for changes in the optimal value function as a function of right-hand side perturbations.

A number of results relating to the differential stability of the optimal value function have also been obtained, generally associated with the existence of directional derivatives or bounds on the directional derivative limit quotient. Danskin [6,7] provided one of the earliest characterizations of the differential stability of the optimal value function of a mathematical program. Addressing the problem minimize $f(x,\varepsilon)$ subject to $x \in S$, S some topological space ε in E^k, Danskin derived conditions under which the directional derivative of f^* exists and also determined its representation. This result, given in Sec. V, Equation (12), has wide applicability in the sense that the constraint space S can be any compact topological space. However, the result is restricted to a constraint set that does not vary with the parameter ε. For the special case in which S is defined by inequalities involving a parameter, $g_i(x,\varepsilon) \geqq 0$ for $i = 1, \ldots, m$, where f is convex and the g_i are concave on S, Hogan [18] has given conditions that imply that the directional derivative of f^* exists and is finite in all directions.

For programs without equality constraints, Rockafellar [24] has shown that, under certain second-order conditions, the optimal value function satisfies a stability of degree two. Under this stability property, bounds on the directional derivative of f^* can be derived. For convex programming problems, Gol'stein [15] has shown that a saddle point condition is satisfied by the directional derivative of f^*. Gauvin and Tolle [14], not assuming convexity, but limiting their analysis to right-hand side perturbations, extended the work of Gol'stein and provide sharp upper and lower bounds on the directional derivative limit quotient of f^*, assuming the Mangasarian-Fromovitz constraint qualification and without requiring the existence of second-order conditions. Sensitivity results for infinite dimensional programs have recently been obtained by Maurer [21,22].

The results reported here extend the work of Gauvin and Tolle to the general inequality constrained mathematical program in which a parameter appears arbitrarily in the constraints and the objective function. Preliminary results for an extension to the general inequality-equality problem are also given. For the inequality problem, we obtain upper and lower bounds on the directional derivative limit quotient of the optimal value function. Though the general strategy of proof given in [11] closely parallels that utilized by Gauvin and Tolle [14], the specific techniques are quite different. It is also relevant to note that the bounds for the inequality-equality problems were obtained essentially simultaneously by Gauvin and Dubeau [13] and also applied by them to show that the optimal value function is locally Lipschitz and to calculate its Clarke generalized derivative. A summary of [13] is contained in these proceedings, in the paper by Gauvin.

II. NOTATION AND DEFINITIONS

In this paper we shall be concerned with mathematical programs of the form:

$$\min_x f(x,\varepsilon) \qquad P(\varepsilon)$$

$$\text{s.t. } g_i(x,\varepsilon) \geqq 0 \ (i = 1,\dots,m), \ h_j(x,\varepsilon) = 0 \ (j = 1,\dots,p)$$

where $x \in E^n$ is the vector of decision variables, ε is a parameter vector in E^k, and the functions f, g_i and h_j are once continuously differentiable on $E^n \times E^k$. The feasible region of problem $P(\varepsilon)$ will be denoted $R(\varepsilon)$ and the set of solutions $S(\varepsilon)$. The m-vector whose components are $g_i(x,\varepsilon)$, $i = 1,\dots,m$, and the p-vector whose components are $h_j(x,\varepsilon)$, $j = 1,\dots,p$, will be denoted by $g(x,\varepsilon)$ and $h(x,\varepsilon)$, respectively.

Following usual conventions the gradient, with respect to x, of a once differentiable real-valued function $f : E^n \times E^k \to E^1$ is denoted $\nabla_x f(x,\varepsilon)$ and is taken to be the row vector $[\partial f(x,\varepsilon)/\partial x_1,\dots,\partial f(x,\varepsilon)/\partial x_n]$. If $g(x,\varepsilon)$ is a vector-valued function, $g : E^n \times E^k \to E^m$, whose components $g_i(x,\varepsilon)$ are differentiable in x, then $\nabla_x g(x,\varepsilon)$ denotes the $m \times n$ Jacobian matrix of g whose ith row is given by $\nabla_x g_i(x,\varepsilon)$, $i = 1,\dots,m$. The transpose of the Jacobian $\nabla_x g(x,\varepsilon)$ will be denoted $\nabla_x' g(x,\varepsilon)$. Differentiation with respect to the vector ε is denoted in a completely analogous fashion.

Henceforth, we do not distinguish between row and column vectors in this paper; their use should be clear from the context in which they are applied.

The Lagrangian for $P(\varepsilon)$ will be written

$$L(x,\mu,\omega,\varepsilon) = f(x,\varepsilon) - \sum_{i=1}^{m} \mu_i g_i(x,\varepsilon) + \sum_{j=1}^{p} \omega_j h_j(x,\varepsilon)$$

and the set of Karush-Kuhn-Tucker vectors corresponding to the decision vector x will be given by

$$K(x,\varepsilon) = \{(\mu,\omega) \in E^m \times E^p : \nabla_x L(x,\mu,\omega,\varepsilon) = 0,$$
$$\mu_i \geqq 0,\ \mu_i g_i(x,\varepsilon) = 0,\ i = 1,\dots,m\}$$

Writing a solution vector as a function of the parameter ε, the index set for inequality constraints which are binding at a solution $x(\varepsilon)$ is denoted by $B(\varepsilon) = \{i : g_i(x(\varepsilon),\varepsilon) = 0\}$. Finally, the optimal value function will be defined as

$$f^*(\varepsilon) = \min \{f(x,\varepsilon) : x \in R(\varepsilon)\}$$

Throughout this paper we shall make use of the well known Mangasarian-Fromovitz Constraint Qualification (MFCQ) which holds at point $x \in R(\varepsilon)$ if: (i) there exists a vector $\tilde{y} \in E^n$ such that

$$\nabla_x g_i(x,\varepsilon)\tilde{y} > 0 \qquad \text{for } i \text{ such that } g_i(x,\varepsilon) = 0 \tag{1}$$

$$\nabla_x h_j(x,\varepsilon)\tilde{y} = 0 \qquad \text{for } j = 1,\dots,p \tag{2}$$

and (ii) the gradients $\nabla_x h_j(x,\varepsilon)$, $j = 1,\dots,p$, are linearly independent.

We shall also have occasion to make use of the notions of semicontinuity for both real-valued functions and point-to-set maps. There are several equivalent definitions for these properties. The ones most suited to our purpose are given below. The reader interested in a more complete development of these properties is referred to Berge [5] and Hogan [20].

DEFINITION 1 Let ϕ be a real-valued function defined on the space X.
(i) ϕ is said to be lower semicontinuous at a point $x_0 \in X$ if

$$\varliminf_{x \to x_0} \phi(x) \geqq \phi(x_0)$$

(ii) ϕ is said to be upper semicontinuous at a point $x_0 \in X$ if

$$\overline{\lim_{x \to x_0}} \ \phi(x) \leqq \phi(x_0)$$

Using these definitions, one readily sees that a real-valued function ϕ is continuous at a point if and only if it is both upper and lower semicontinuous at that point.

DEFINITION 2 Let $\phi : X \to Y$ be a point-to-set mapping and let $\{\varepsilon_n\} \subset X$ with $\varepsilon_n \to \bar{\varepsilon}$ ($\bar{\varepsilon}$ in X). (i) ϕ is said to be lower semicontinuous at a point $\bar{\varepsilon}$ of X if, for each $\bar{x} \in \phi(\bar{\varepsilon})$, there exists a value n_0 and a sequence $\{x_n\} \subset Y$ with $x_n \in \phi(\varepsilon_n)$ for $n \geqq n_0$ and $x_n \to \bar{x}$. (ii) ϕ is said to be upper semicontinuous at a point $\bar{\varepsilon}$ of X if $x_n \in \phi(\varepsilon_n)$ and $x_n \to \bar{x}$ together imply that $\bar{x} \in \phi(\bar{\varepsilon})$.

Following Berge [5], we denote the lower (upper) semicontinuity of point-to-set maps by l.s.c. (u.s.c.); for real-valued functions we use the notation lsc and usc for lower and upper semicontinuity, respectively.

DEFINITION 3 A point-to-set mapping $\phi : X \to Y$ is said to be uniformly compact near a point $\bar{\varepsilon}$ of X if the closure of the set $\cup\phi(\varepsilon)$ for ε in $N(\bar{\varepsilon})$ is compact for some neighborhood $N(\bar{\varepsilon})$ of $\bar{\varepsilon}$.

In Sec. III, we apply a reduction of variables technique to $P(\varepsilon)$ which transforms that program to an equivalent program involving only inequality constraints. This approach simplifies the derivation of intermediate results which are needed to derive the bounds on the directional derivative limit quotients of $f^*(\varepsilon)$ given in Sec. IV. Sec. V concludes with a few remarks concerning related results.

III. REDUCTION OF VARIABLES

Since MFCQ is assumed, the number p of equality constraints must be less than n. If problem $P(\varepsilon)$ contains no equality constraints, then reference to them should be ignored in the development which follows. However, in the event that equalities are present, the linear independence criterion of MFCQ will be invoked and, by means of an implicit function theorem, the equality constraints of $P(\varepsilon)$ will be eliminated.

Specifically, under the given conditions and possibly after relabeling the variables, if $h(x^*,\varepsilon^*) = 0$ then there exists a unique continuous function $x_D = x_D(x_I,\varepsilon)$ which satisfies the system $h(x_D,x_I,\varepsilon) = 0$ identically in a neighborhood of the point $(x_D^*,x_I^*,\varepsilon^*) \in E^p \times E^{n-p} \times E^k$. Moreover, since we have assumed that the mapping h is once continuously differentiable in x_I and ε, the function $x_D(x_I,\varepsilon)$ is also. Thus, near the point (x_I^*,ε^*), we have that $h(x_D(x_I,\varepsilon),x_I,\varepsilon) \equiv 0$, so that writing $x = (x_D,x_I)$ and substituting $x_D(x_I,\varepsilon)$ for x_D in the program $P(\varepsilon)$, we obtain a program involving only inequality constraints. Letting $\tilde{f}(x_I,\varepsilon) = f[x_D(x_I,\varepsilon),x_I,\varepsilon]$ and $\tilde{g}(x_I,\varepsilon) = g[x_D(x_I,\varepsilon),x_I,\varepsilon]$, $P(\varepsilon)$ is thus reduced to the problem

$$\min_{x_I} \tilde{f}(x_I,\varepsilon) \qquad \tilde{P}(\varepsilon)$$

$$\text{s.t.} \quad \tilde{g}_i(x_I,\varepsilon) \geqq 0 \qquad (i = 1,\ldots,m)$$

The local equivalence of the problems $P(\varepsilon)$ and $\tilde{P}(\varepsilon)$ is summarized by Lemma 1, a consequence of the 1-1 transformation induced by the implicit function theorem. For notational simplicity and without loss of generality we assume that $\varepsilon^* = 0$.

LEMMA 1 If $f,g,h \in C^1$, and the once continuously differentiable vector function $x_D = x_D(x_I,\varepsilon)$ is given such that $h(x_D(x_I,\varepsilon),x_I,\varepsilon) \equiv 0$ in a neighborhood of $(x^*,0) = (x_D^*,x_I^*,0)$, then near $\varepsilon = 0$, the point $x(\varepsilon)$ satisfies the Karush-Kuhn-Tucker first-order necessary conditions for an optimum of $P(\varepsilon)$ if and only if the point $x_I(\varepsilon)$ is a Karush-Kuhn-Tucker point of $\tilde{P}(\varepsilon)$. Furthermore, near $\varepsilon = 0$, $x(\varepsilon)$ is a local solution of $P(\varepsilon)$ if and only if $x_I(\varepsilon)$ is a local solution of $\tilde{P}(\varepsilon)$, where $x(\varepsilon) = (x_D(x_I(\varepsilon),\varepsilon),x_I(\varepsilon))$.

We now turn to the development of the results for the general inequality constrained problem $\tilde{P}(\varepsilon)$. Proofs of the subsidiary results are omitted when they are felt to be straightforward. All proofs and further details may be found in [11]. We first note in passing that the Mangasarian-Fromovitz constraint qualification for $P(\varepsilon)$ is inherited by the reduced program $\tilde{P}(\varepsilon)$, in a very well specified and constructive manner. This result is clearly relevant to relating any results obtained for $\tilde{P}(\varepsilon)$ under MFCQ to $P(\varepsilon)$ under MFCQ, although this extension will not be completed here.

THEOREM 1 If $f,g,h \in C^1$, then MFCQ holds for $P(0)$ at $(x^*,0) = (x_D^*,x_I^*,0)$ with $\tilde{y} = (\tilde{y}_D,\tilde{y}_I) \in E^n$ the associated vector, where $\tilde{y}_D \in E^p$ and $\tilde{y}_I \in E^{n-p}$, if and only if MFCQ holds in problem $\tilde{P}(0)$ at the point $(x_I^*,0)$ with the vector $\tilde{y}_I$.

The proof follows by using the indicated partition, and differentiating (2) and $h(x,\varepsilon) = 0$, with $x_D = x_D(x_I,\varepsilon)$.

The set of Karush-Kuhn-Tucker multipliers corresponding to a solution, x^*, of $P(0)$ has been shown by Gauvin and Tolle [14] to be nonempty, compact, and convex if and only if MFCQ is satisfied at x^*. The use of that result in the proof of the next theorem permits the derivation of the relationships (4) and (6) which lead directly, in the next section, to the establishment of the upper and lower bounds on $D_z\tilde{f}^*$. In particular, as will become evident, (4) and (6) show that if MFCQ is satisfied at a local solution of $\tilde{P}(0)$, then from that solution, there is a direction in which the directional derivative, with respect to x, of the objective function yields that portion of the bound attributable to the constraint perturbation. The result given in (3) is needed subsequently to show that $\tilde{P}(\varepsilon)$ has feasible points for all ε near 0.

For notational simplicity, throughout the remainder of this paper (unless otherwise specified) we shall refer to the problem functions of the program $\tilde{P}(\varepsilon)$ without the "tilde" notation, and without reference to the variable reduction derivation. That is, assume that an inequality constrained problem of the form $\tilde{P}(\varepsilon)$ is given. The distinction between reference to $P(\varepsilon)$ and $\tilde{P}(\varepsilon)$ will be clear from the text. Also, the decision variables for the reduced problem will not be subscripted, and, unless specified to the contrary, all gradients will be understood to be taken with respect to the relevant decision variable. The following theorem and proof follow the constructs used by Gauvin and Tolle [14] for the right hand side perturbation problem. It is emphasized that the given results for $\tilde{P}(\varepsilon)$ are valid for the general inequality constrained problem, and taken in this context, are independent of and not related to the problem inherited from an inequality-equality constrained problem.

THEOREM 2 If the conditions of MFCQ are satisfied for some $x^*\varepsilon\tilde{S}(0)$, then, for any direction $z\varepsilon E^k$, there exists a vector $\bar{y}\varepsilon E^{n-p}$ satisfying:

$$\text{(i)} \quad -\nabla g_i(x^*,0)\bar{y} \leqq z\nabla_\varepsilon g_i(x^*,0) \text{ for } i\varepsilon\tilde{B}(0) \tag{3}$$

$$\text{(ii)} \quad \nabla f(x^*,0)\bar{y} = \max_{\mu\varepsilon\tilde{K}(x^*,0)} [-z\nabla'_\varepsilon g(x^*,0)\mu] \tag{4}$$

PROOF. Given $z \epsilon E^k$, consider the following linear program:

$$\max_{\mu} \quad [-z\nabla'_{\varepsilon} g(x^*,0)\mu]$$

$$\text{s.t.} \quad \mu\nabla g(x^*,0) = \nabla f(x^*,0)$$

$$\mu_i g_i(x^*,0) = 0 \qquad (i = 1,\ldots,m)$$

$$\mu_i \geqq 0 \qquad (i = 1,\ldots,m)$$

Since MFCQ is assumed to hold at $(x^*,0)$, from [14] we have that $\tilde{K}(x^*,0)$ is nonempty, compact and convex. Thus the program above is bounded and feasible. The result now follows by invoking the duality theorem of linear programming.

If we replace z by -z in Theorem 2, then we have the following obvious realization of Theorem 2, which we restate as a corollary simply for convenience and clarity. It is used to prove Theorem 5.

COROLLARY 1 If the conditions of MFCQ are satisfied for some $x^* \epsilon \tilde{S}(0)$, then for any direction $z \epsilon E^k$, there exists a vector $\bar{y} \epsilon E^{n-p}$ satisfying:

$$\text{(i)} \quad \nabla g_i(x^*,0)\bar{y} \geqq z\nabla_{\varepsilon} g_i(x^*,0) \text{ for } i \epsilon \tilde{B}(0) \tag{5}$$

$$\text{(ii)} \quad \nabla f(x^*,0)\bar{y} = \max_{\mu \epsilon \tilde{K}(x^*,0)} [z\nabla'_{\varepsilon} g(x^*,0)\mu] \tag{6}$$

In order to establish the bounds given in the next section, we require the next result, which shows that $\tilde{P}(\varepsilon)$ has points of feasibility along any ray emanating from $\varepsilon = 0$.

THEOREM 3 Let $g: E^{n-p} x E^k \to E^m$, with $g \epsilon C^1$. If MFCQ holds at $x^* \epsilon \tilde{R}(0)$ then, for any unit vector $z \epsilon E^k$ and any $\delta > 0$, $g(x^* + \beta(\bar{y} + \delta\tilde{y}), \beta z) > 0$ for β positive and sufficiently near zero, where $\bar{y}$ is any vector satisfying (3).

This result is easily proved by using a first order Taylor expansion of $g_i(x^* + \beta(\bar{y} + \delta\tilde{y}), \beta z)$ about $(x^*,0)$, along with (1) and (3).

The next theorem, which is needed in the proof of Theorem 7, also implies (in conjunction with Theorem 3) the interesting fact that $\tilde{R}(\varepsilon)$, the feasible region for $\tilde{P}(\varepsilon)$, and $\tilde{R}(0)$, the feasible region for $\tilde{P}(0)$, have points in common when ε is sufficiently near zero. It says essentially that any sequence $\{x_n\}$ of points extracted respectively from $\tilde{R}(\varepsilon_n)$, and converging to an MFCQ point of $\tilde{R}(0)$, must yield a sequence of points in $\tilde{R}(0)$ when displaced in a given specified direction.

THEOREM 4 Let $\beta_n \to 0^+$ in E^1, let z be any unit vector in E^k, and let $\delta > 0$. If $x_n \varepsilon \tilde{R}(\beta_n z)$, with $x_n \to x^* \varepsilon \tilde{R}(0)$, and if the conditions of MFCQ are satisfied at x^*, then $x_n + \beta_n(\bar{y} + \delta\tilde{y}) \varepsilon \tilde{R}(0)$ for n sufficiently large, where $\bar{y}$ satisfies (5) and $\tilde{y}$ is given by the constraint qualification.

In the proof of Theorem 7 we shall need both the upper semicontinuity of $\tilde{R}(\varepsilon)$, which follows readily, and the continuity of $f^*(\varepsilon)$. Gauvin and Dubeau [13] and Fiacco [9] proved that the hypotheses used in the current paper are also sufficient for the latter results.

LEMMA 2 If $\tilde{R}(0)$ is nonempty and $\tilde{R}(\varepsilon)$ is uniformly compact for ε near zero, then $\tilde{R}(\varepsilon)$ is a u.s.c. mapping at $\varepsilon = 0$.

THEOREM 5 If $\tilde{R}(0)$ is nonempty, $\tilde{R}(\varepsilon)$ is uniformly compact for ε near zero, and if the conditions of MFCQ hold at $x^* \varepsilon \tilde{S}(0)$, then $f^*(\varepsilon)$ is continuous at $\varepsilon = 0$.

IV. BOUNDS ON THE PARAMETRIC VARIATION OF THE OPTIMAL VALUE FUNCTION

Given these preliminary results, the desired bounds are readily obtained. We state and prove the main results.

THEOREM 6 If, for $\tilde{P}(\varepsilon)$, MFCQ holds for some $x^* \varepsilon \tilde{S}(0)$, then for any direction $z \varepsilon E^k$,

$$\limsup_{\beta \to 0^+} \frac{f^*(\beta z) - f^*(0)}{\beta} \leq \max_{\mu \varepsilon \tilde{K}(x^*,0)} z \nabla_\varepsilon \tilde{L}(x^*,\mu,0) \tag{7}$$

PROOF. Let $\bar{y}$ satisfy (3) and (4), let $\tilde{y}$ be given by the constraint qualification, and let β satisfy the conditions of Theorem 3. Then for any $\delta > 0$ and any $z \varepsilon E^k$, there exists a value $\bar{\beta} > 0$ such that $x^* + \beta(\bar{y} + \delta\tilde{y}) \varepsilon \tilde{R}(\beta z)$ for all $\beta \varepsilon (0,\bar{\beta}]$. Thus

$$\limsup_{\beta \to 0^+} \frac{f^*(\beta z) - f^*(0)}{\beta} \leq \frac{df}{d\beta}(x^*,0) = (\bar{y} + \delta\tilde{y})\nabla f(x^*,0) + z\nabla_\varepsilon f(x^*,0)$$

The result now follows by noting that the inequality above holds for arbitrary positive δ (so the limit as $\delta \to 0$ can be taken) and by then applying (4).

COROLLARY 1 Under the hypotheses of the previous theorem, if MFCQ holds at each point $x \varepsilon \tilde{S}(0)$, then

$$\limsup_{\beta\to 0^+} \frac{f^*(\beta z) - f^*(0)}{\beta} \leqq \inf_{x\varepsilon\tilde{S}(0)} \max_{\mu\varepsilon\tilde{K}(x,0)} z\nabla_\varepsilon \tilde{L}(x,\mu,0) \tag{8}$$

PROOF. The result follows directly by applying the previous theorem at each point of $\tilde{S}(0)$.

THEOREM 7 If, for $\tilde{P}(\varepsilon)$, $\tilde{R}(0)$ is nonempty, $\tilde{R}(\varepsilon)$ is uniformly compact near $\varepsilon = 0$, and MFCQ holds for each $x\varepsilon S(0)$, then, for any direction $z\varepsilon E^k$, z a unit vector,

$$\liminf_{\beta\to 0^+} \frac{f^*(\beta z) - f^*(0)}{\beta} \geqq \min_{\mu\varepsilon\tilde{K}(x^*,0)} z\nabla_\varepsilon \tilde{L}(x^*,\mu,0) \tag{9}$$

holds for some $x^*\varepsilon\tilde{S}(0)$.

PROOF. Let $x_n\varepsilon\tilde{S}(\beta_n z)$ and let $\beta_n\to 0^+$ be such that

$$\liminf_{\beta\to 0^+} \frac{f^*(\beta z) - f^*(0)}{\beta} = \lim_{n\to\infty} \frac{f(x_n,\beta_n z) - f(x^*,0)}{\beta_n}$$

Since $R(\varepsilon)$ is uniformly compact we can choose a subsequence, which we again denote by $\{x_n\}$, and a vector x^* such that $x_n\to x^*$.

Using Lemma 2 and Theorem 5 it follows that $x^*\varepsilon\tilde{S}(0)$. By Theorem 4, $x_n + \beta_n(\bar{y} + \delta\tilde{y})\ \varepsilon\ \tilde{R}(0)$ for n sufficiently large, so that an application of the mean value theorem yields

$$\liminf_{\beta\to 0^+} \frac{f^*(\beta z) - f^*(0)}{\beta} \geqq \lim_{n\to\infty} [-(\bar{y} + \delta\tilde{y})\nabla f(\alpha_n) + z\nabla_\varepsilon f(\alpha_n)]$$

where $\alpha_n\varepsilon E^{n-p} \times E^k$ is given by the mean value theorem.

Passing to the limit in the above inequality, noting that δ was chosen as any positive value, and applying (6), we may conclude that for any $x^*\varepsilon\tilde{S}(0)$ where MFCQ holds,

$$\liminf_{\beta\to 0^+} \frac{f^*(\beta z) - f^*(0)}{\beta} \geqq \min_{\mu\varepsilon\tilde{K}(x^*,0)} z\nabla_\varepsilon \tilde{L}(x^*,\mu,0)$$

The next corollary, although it gives a bound that is weaker than the one just derived, is useful in that it permits us, in Corollary 4, to demonstrate an instance in which $D_z f^*(0)$ can be shown to exist.

COROLLARY 3 Under the hypotheses of the previous theorem

$$\liminf_{\beta \to 0^+} \frac{f^*(\beta z) - f^*(0)}{\beta} \geq \inf_{x \in \tilde{S}(0)} \min_{\mu \in \tilde{K}(x,0)} z \nabla_\varepsilon \tilde{L}(x,\mu,0) \tag{10}$$

Under the reduction of variables that was applied above one readily obtains that, in a neighborhood of $(x_D(x_I^*,0),x_I^*,0) = (x^*,0)$, $L(x,\mu,\omega,\varepsilon) \equiv \tilde{L}(x_I,\mu,\varepsilon)$, and that with $\omega = -(\nabla_{x_D} f - \mu \nabla'_{x_D} g)(\nabla'_{x_D} h^{-1})$, $\nabla_\varepsilon \tilde{L} = \nabla_\varepsilon L$. The results obtained above for $\tilde{P}(\varepsilon)$ can thus readily be related, locally, to $P(\varepsilon)$.

Returning to $\tilde{P}(\varepsilon)$, since the uniqueness of the Karush-Kuhn-Tucker multipliers corresponding to any particular point $x^* \varepsilon \tilde{S}(0)$ is guaranteed by the linear independence of the binding constraint gradients, the next corollary follows directly from Corollaries 1 and 3.

COROLLARY 4 Assume $\tilde{R}(0)$ is nonempty and $\tilde{R}(\varepsilon)$ is uniformly compact near $\varepsilon = 0$. If the gradients, taken with respect to x, of the constraints binding at x^* are linearly independent for each $x^* \varepsilon \tilde{S}(0)$, then for any unit vector $z \varepsilon E^k$, $D_z f^*(0)$ exists and is given by

$$D_z f^*(0) = \inf_{x \varepsilon S(0)} z \nabla_\varepsilon L(x,\mu(x,0),0)$$

where $(\mu(x,0))$ is the unique multiplier vector associated with x.

The next theorem asserts that the directional derivative $D_z f^*(0)$ exists when $P(\varepsilon)$ is a convex program and its form is immediately obtained from the prior results. The convexity of the Lagrangian in this instance leads to the equality of the upper and lower bounds given by (8) and (10). The result follows for $\tilde{P}(\varepsilon)$ and is inherited by the inequality-equality problem $P(\varepsilon)$ because of Theorem 1 and the fact that the assumptions are strong enough to guarantee that the variable reduction transformation is valid globally; hence $\tilde{P}(\varepsilon)$ and $P(\varepsilon)$ are equivalent globally, in this instance.

THEOREM 8 In $P(\varepsilon)$, let $f(x,\varepsilon)$ and $-g_i(x,\varepsilon)$, $i = 1,\ldots,m$ be convex and let $h_j(x,\varepsilon)$, $j = 1,\ldots,p$ be affine in x. If $R(0)$ is nonempty, $R(\varepsilon)$ is uniformly compact near $\varepsilon = 0$, and MFCQ holds for each $x^* \varepsilon S(0)$, then, for any unit vector $z \varepsilon E^k$,

$$D_z f^*(0) = \inf_{x \varepsilon S(0)} \max_{(\mu,\omega) \varepsilon K(x,0)} z \nabla_\varepsilon L(x,\mu,\omega,0) \tag{11}$$

V. RELATED RESULTS

Danskin [6,7] provided a now well-known characterization of the directional derivative of the optimal value function of $P(\varepsilon)$ in the case that the constraints are independent of a parameter. Under the conditions that the region of feasibility, $R(0)$, is compact, and $f(x,\varepsilon)$ and $\nabla_\varepsilon f(x,\varepsilon)$ are continuous at $\varepsilon = 0$, Danskin showed that

$$D_z f^*(0) = \min_{x \in S(0)} z \nabla_\varepsilon f(x,0) \tag{12}$$

Relating our hypotheses to Danskin's construct, we first note the equivalence of our assumption of the uniform compactness of $R(\varepsilon)$ for ε near $\varepsilon = 0$, and the assumption that the feasible region is compact if the constraints of $P(\varepsilon)$ do not depend on ε. To see this, one need only consider that, in this case, $R(\varepsilon) \equiv R(0)$ for all ε and apply Definition 3. In addition, when the feasible region is independent of ε, the proofs of Theorems 6 and 7 remain valid by simply suppressing all reference to the dependence of the constraints on ε and by considering the unperturbed point x^* instead of $x^* + \beta(\bar{y} + \delta\tilde{y})$. One is then led to conclude that

$$\limsup_{\beta \to 0^+} \frac{f^*(\beta z) - f^*(0)}{\beta} \leqq \min_{x \in S(0)} z \nabla_\varepsilon f(x,0) \tag{13}$$

$$\liminf_{\beta \to 0^+} \frac{f^*(\beta z) - f^*(0)}{\beta} \geqq \min_{x \in S(0)} z \nabla_\varepsilon f(x,0) \tag{14}$$

for any unit vector $z \in E^k$. Thus it follows that, under the stated conditions, namely the compactness of R and the continuity of $f(x,\varepsilon)$ and $\nabla_\varepsilon f(x,\varepsilon)$ at $\varepsilon = 0$, our results are consistent with those of Danskin in that they verify the existence of $D_z f^*(0)$ and show [from (13) and (14)] that it can be expressed as in (12).

As indicated, Gauvin and Tolle [14] obtained analogous results for the programs with right-hand side perturbations, i.e., for programs of the form

$$\begin{aligned} &\min \quad f(x) \\ &\text{s.t.} \quad g_i(x) \geqq \varepsilon_i \qquad (i = 1,\dots,m) \\ &\qquad\;\; h_j(x) = \varepsilon_{m+j} \qquad (j = 1,\dots,p) \end{aligned} \tag{$P'(\varepsilon)$}$$

The bounds we have given for the inequality program $\tilde{P}(\varepsilon)$ reduce to those obtained in [14] for the more restrictive perturbations appearing in $P'(\varepsilon)$ when the equalities are absent, and the directional derivative obtained for the general convex program $P(\varepsilon)$ (Theorem 8) extends the analogous result obtained in [14] for $P'(\varepsilon)$.

The existence of $D_z f^*(0)$ assured by Theorem 8, and its expression (11), corresponds under slightly different assumptions, with results obtained for convex programs by Gol'stein [15] and Hogan [18]. Our result is a direct extension to the general perturbed convex mathematical program of a result given by Gauvin and Tolle [14] for convex right-hand side programs.

Auslender [4] has extended the results of Gauvin and Tolle [14] to problems involving nondifferentiable functions. In particular, the bounds given in [14] are obtained for right-hand side programs in which the problem functions are locally Lipschitz and those defining the quality constraints are continuously differentiable.

After completing this paper, we learned that Geraud Fontanie ("Locally Lipschitz Functions and Nondifferentiable Programming," M.S. Thesis, Technical Report 80-3, Curriculum in Operations Research and Systems Analysis, University of North Carolina at Chapel Hill, 1980) extended the Gauvin-Tolle bounds [14] to a generally perturbed Lipschitz program, using the reduction technique described here and first employed in [11] in this context.

Finally, we note that the full extension of the bounds results reported here, to the general finite dimensional inequality-equality constrained once differentiable program, using the reduction technique, has been recently obtained by Fiacco.

REFERENCES

1. R. L. Armacost and A. V. Fiacco, NLP sensitivity analysis for RHS perturbations: A brief survey and second order extensions, Technical Paper T-334, Institute for Management Science and Engineering, The George Washington University, April 1976.

2. R. L. Armacost and A. V. Fiacco, Second-order parametric sensitivity analysis in NLP and estimates by penalty function methods, Technical Paper T-324, Institute for Management Science and Engineering, The George Washington University, December 1975.

3. R. L. Armacost and A. V. Fiacco, Sensitivity analysis for parametric nonlinear programming using penalty methods, *Computers and Mathematical Programming*, National Bureau of Standards Special Publication 502 (1978), 261-269.

4. A. Auslender, Differential stability in nonconvex and nondifferentiable programming, *Mathematical Programming Study 10*(1979), 29-41.

5. C. Berge, *Topological Spaces*, trans. by E. M. Patterson, New York: Macmillan, 1963.

6. J. M. Danskin, *The Theory of Max-Min*, New York: Springer-Verlag, 1967.

7. J. M. Danskin, The theory of max-min with applications, *SIAM J. Appl. Math. 14*(1966), 641-665.

8. J. P. Evans and F. J. Gould, Stability in nonlinear programming, *Operations Res. 18*(1970), 107-118.

9. A. V. Fiacco, Continuity of the optimal value function under the Mangasarian-Fromovitz constraint qualification, Technical Paper T-432, Institute for Management Science and Engineering, The George Washington University, September 1980.

10. A. V. Fiacco, Sensitivity analysis for nonlinear programming using penalty methods, *Math. Programming 10*(1976), 287-311.

11. A. V. Fiacco and W. P. Hutzler, Extensions of the Gauvin-Tolle optimal value differential stability results to general mathematical programs, Technical Paper T-393, Institute for Management Science and Engineering, The George Washington University, April 1979.

12. A. V. Fiacco and G. P. McCormick, *Nonlinear Programming: Sequential Unconstrained Minimization Techniques*, New York: Wiley, 1968.

13. J. Gauvin and F. Dubeau, Differential properties of the marginal function in mathematical programming, Technical Report, Ecole Polytechnique de Montréal, June 1979.

14. J. Gauvin and J. W. Tolle, Differential stability in nonlinear programming, *SIAM J. Control and Optimization 15*(1977), 294-311.

15. E. G. Gol'stein, *Theory of Convex Programming*, Translations of Mathematical Monographs, Vol. 36, Providence: American Mathematical Society, 1972.

16. H. J. Greenberg and W. P. Pierskalla, Extensions of the Evans-Gould stability theorems for mathematical programs, *Operations Res. 20* (1972), 143-153.

17. W. Hogan, The continuity of the perturbation function of a convex program, *Operations Res. 21*(1973), 351-352.

18. W. Hogan, Directional derivatives for extremal value functions with applications to the completely convex case, *Operations Res. 21*(1973), 188-209.

19. W. Hogan, Optimization and convergence for extremal value functions arising from structured nonlinear programs, Doctoral Dissertation, Western Management Science Institute, University of California at Los Angeles, September 1971.

20. W. Hogan, Point-to-set maps in mathematical programming, *SIAM Review 15*(1973), 591-603.

21. H. Maurer, A sensitivity result for infinite nonlinear programming problems; Part I: Theory, Mathematisches Institut der Universität Würzburg, Würzburg, W. Germany, January 1977.

22. H. Maurer, A sensitivity result for infinite nonlinear programming problems; Part II: Applications to optimal control problems, Mathematisches Institut der Universität Würzburg, Würzburg, W. Germany, January 1977.

23. S. M. Robinson, Perturbed Kuhn-Tucker points and rates of convergence for a class of nonlinear programming algorithms, *Math. Programming 7* (1974), 1-16.

24. R. T. Rockafellar, Augmented Lagrange multiplier functions and duality in nonconvex programming, *SIAM J. Control and Optimization 12* (1974), 268-285.

Chapter 4 THE GENERALIZED GRADIENT OF THE PERTURBATION FUNCTION IN MATHEMATICAL PROGRAMMING

JACQUES GAUVIN / Ecole Polytechnique de Montréal, Montréal, Québec, Canada

ABSTRACT

Estimates are given for the Clarke generalized gradient of the perturbation function of a nonlinear mathematical program with equality and inequality constraints where a perturbation vector appears in all functions defining the program.

In this short communication estimates are given for the Clarke generalized gradient of the perturbation or marginal function of a mathematical program where a perturbation or parameter vector is present.

The nonlinear mathematical program considered is

$$\begin{aligned} &\max f(x,y),\ x \in \mathbb{R}^n,\ y \in \mathbb{R}^m \\ &\text{subject to } g_i(x,y) \leqq 0,\ i = 1,\ldots,p \\ &\qquad\qquad h_j(x,y) = 0,\ j = 1,\ldots,q \end{aligned} \tag{1}$$

where the components of the vector y can be seen as parameters or perturbations. For each value of y, the set of feasible solutions is $S(y) = \{x \mid g_i(x,y) \leqq 0,\ h_j(x,y) = 0,\ i = 1,\ldots,p,\ j = 1,\ldots,q\}$. The marginal or perturbation function is the optimal value of the program $v(y) = \sup\{f(x,y) | x \in S(y)\}$ and the set of optimal solutions is $P(y) = \{\bar{x} \in S(y) | f(\bar{x},y) = v(y)\}$.

The Lagrangian function corresponding to program (1) is

$$L(x,y;u,v) = f(x,y) - \sum_{i=1}^{p} u_i g_i(x,y) - \sum_{j=1}^{q} w_j h_j(x,y)$$

where all functions are assumed continuously differentiable. For an optimal point $\bar{x} \in P(\bar{y})$, $K(\bar{x},\bar{y})$ is the set of Kuhn-Tucker vectors associated with $\bar{x}$; that is, the set of $(u,w) \in \mathbb{R}^p \times \mathbb{R}^q$ such that

(i) $\nabla_x L(\bar{x},\bar{y};u,w) = 0$

(ii) $u_i g_i(\bar{x},\bar{y}) = 0$

(iii) $u_i \geqq 0 \qquad i = 1,\ldots,p$

where ∇_x denotes the gradient relative to the variable x.

If $\bar{x}$ is a local maximum of program (1), the Mangasarian-Fromovitz regularity condition is known to be necessary and sufficient to have $K(\bar{x},\bar{y})$ *nonempty* and *bounded* (Gauvin [3]). This regularity condition (M-F) is:

(i) There exists a direction $\bar{r} \in \mathbb{R}^n$ such that $\nabla_x g_i(\bar{x},\bar{y})\bar{r} < 0$, $i \in I(\bar{x},\bar{y}) = \{i | g_i(\bar{x},\bar{y}) = 0\}$; $\nabla_x h_j(\bar{x},\bar{y})\bar{r} = 0$, $j = 1,\ldots,q$.

(ii) The gradients $\{\nabla_x h_j(\bar{x},\bar{y}),\ j = 1,\ldots,q\}$ are linearly independent.

In order to present an outline of the proof of the results of this paper, the following definitions are needed.

The *lower* and *upper Dini directional derivatives* of v(y) at $\bar{y}$ in the direction $s \in \mathbb{R}^m$ are respectively

$$D_+ v(\bar{y};s) = \liminf_{t \to 0^+} [v(\bar{y} + ts) - v(\bar{y})]/t$$

$$D^+ v(\bar{y};s) = \limsup_{t \to 0^+} [v(\bar{y} + ts) - v(\bar{y})]/t$$

The marginal function v(y) is *locally Lipschitz* near $\bar{y}$ if for some neighborhood $N(\bar{y})$ of $\bar{y}$ there exists an $M > 0$ such that for any $y_1,y_2 \in N(\bar{y})$

$$|v(y_2) - v(y_1)| \leq M||y_2 - y_1||$$

A locally Lipschitz function on $N(\bar{y})$ has, at almost all $y \in N(\bar{y})$, a gradient $\nabla v(y)$.

The *Clarke generalized gradient* of v(y) at $\bar{y}$, denoted $\partial v(y)$, is the convex hull of the set $\{\lim \nabla v(y_n)\}$ of limits, where $\nabla v(y_n)$ exists and

$y_n \to \bar{y}$ as $n \to \infty$; $\partial v(\bar{y})$ is a nonempty convex compact set (Clarke [1]).

The *Clarke generalized directional derivative* of $v(y)$ at $\bar{y}$ in the direction s is

$$D^o v(\bar{y};s) = \limsup_{\substack{y \to \bar{y} \\ t \to 0^+}} [v(y + tx) - v(y)]/t$$

If $v(y)$ is locally Lipschitz near $\bar{y}$, then, for any direction $s \in \mathbb{R}^n$, $D^o v(\bar{y};s)$ is the support function of $\partial v(\bar{y})$; that is [1],

$$D^o v(\bar{y};s) = \max \{\xi \cdot s \mid \xi \in \partial v(\bar{y})\}$$

The function $v(y)$ is said to be *Clarke regular at* $\bar{y}$ if the ordinary directional derivative $Dv(\bar{y};s)$ exists and is equal to $D^o v(\bar{y};s)$ for any direction $s \in \mathbb{R}^m$.

The next two results give a characterization of the Clarke generalized gradient of the perturbation function $v(y)$ at some point $\bar{y}$ where $v(\bar{y})$ is finite.

THEOREM 1 Suppose the feasible set $S(\bar{y})$ is nonempty, $S(y)$ is uniformly compact near $\bar{y}$, and the (M-F) regularity condition holds at every optimal point $\bar{x} \in P(\bar{y})$. Then the Clarke generalized gradient of $v(y)$ at $\bar{y}$ is included in the convex hull of the gradients $\nabla_y L(\bar{x},\bar{y};u,v)$ for $\bar{x} \in P(\bar{y})$, $(u,v) \in K(\bar{x},\bar{y})$; that is,

$$\partial v(\bar{y}) \subseteq \text{co}\left\{ \bigcup_{\bar{x} \in P(\bar{y})} \ \bigcup_{(u,v) \in K(\bar{x},\bar{y})} \nabla_y L(\bar{x},\bar{y};u,v) \right\}$$

OUTLINE OF THE PROOF. This theorem is obtained after a sequence of results developed in Gauvin-Dubeau [5]. First, with $S(\bar{y})$ nonempty and $S(y)$ uniformly compact year $\bar{y}$ (i.e., the closure of $\bigcup S(y)$, y near $\bar{y}$, is compact), the marginal function $v(y)$ is upper semicontinuous at $\bar{y}$ (Lemma 3.2 in [5]); and if at least for one $\bar{x} \in P(\bar{y})$, the (M-F) regularity condition holds, then $v(y)$ is also lower semicontinuous at $\bar{y}$ (Theorem 3.3 in [5]); therefore, under the assumptions of the theorem $v(y)$ is continuous at $\bar{y}$. From Corollary 4.3 in [5], the lower and upper Dini directional derivatives are bounded in the following manner:

$$\begin{aligned} \sup_{\bar{x} \in P(\bar{y})} \ \min_{(u,v) \in K(\bar{x},\bar{y})} \{\nabla_y L(\bar{x},\bar{y});u,v)s\} &\leqq D_+ v(\bar{y};s) \leqq D^+ v(\bar{y};s) \\ &\leqq \sup_{\bar{x} \in P(\bar{y})} \ \max_{(u,v) \in K(\bar{x},\bar{y})} \{\nabla_y L(\bar{x},\bar{y};u,v)s\} \end{aligned}$$

This result has also been obtained recently by Fiacco and Hutzler [2] using an approach different from the one in [5]. From Theorem 3.7 in [5], the (M-F) regularity condition is preserved for any optimal point $\bar{x} \in P(y)$, for y near $\bar{y}$ (see Robinson [6] for a similar result), and the set $K(y) = \bigcup_{\bar{x} \in P(y)} K(\bar{x};y)$ is uniformly compact near $\bar{y}$. Therefore the Dini directional derivatives are uniformly bounded near $\bar{y}$. From this it can be obtained easily that v(y) is locally Lipschitz at $\bar{y}$ (Theorem 5.1 in [5]). The result is then obtained (Theorem 5.3 in [5]). □

A known example (see Example 1 in [4]) shows that the perturbation function v(y) is not necessarily Clarke regular at $\bar{y}$ under the (M-F) regularity condition. But if we assume the more restrictive linear independence (L.I.) regularity condition: The gradients $\{\nabla_x g_i(\bar{x},\bar{y}),\ i \in I(\bar{x},\bar{y}),\ \nabla_x h_j(\bar{x},\bar{y}),\ j = 1,\ldots,q\}$ are linearly independent (L.I.), then we have the following sharp result (Corollary 5.4 in [5]).

COROLLARY 1 If the hypotheses of Theorem 1 hold with the regularity condition (L.I.) instead of (M-F), then the perturbation function v(y) is Clarke regular at $\bar{y}$ and its Clarke generalized gradient is given by

$$\partial v(\bar{y}) = \mathrm{co}\left\{ \bigcup_{\bar{x} \in P(\bar{y})} \nabla_x L(\bar{x},\bar{y};\bar{u},\bar{w}) \right\}$$

where $(\bar{u},\bar{w})$ is the unique multiplier vector corresponding to $\bar{x}$.

The results of this communication are a generalization to general perturbations of similar results in Gauvin [4] obtained for the special and simpler case of right-hand perturbation:

$$\begin{aligned} &\max f(x),\ x \in \mathbb{R}^n \\ &\text{subject to } g_i(x) \leqq y_i,\ i = 1,\ldots,p \\ &\qquad\qquad\ h_j(x) = y_j,\ j = 1,\ldots,q \end{aligned}$$

These results can also be seen as a generalization to nonconvex mathematical programming of a well known result of convex programming wherein the subdifferential of the marginal or perturbation function is equal to the set of Kuhn-Tucker vectors (see Theorem 29.1 in Rockafellar [7]). Since, under the convexity assumption, the Clarke generalized gradient reduces to the subdifferential, the generalization can be considered satisfactory.

REFERENCES

1. F. H. Clarke, Generalized gradients and applications, *Trans. Amer. Math. Soc. 205*(1975), 247-262.

2. A. V. Fiacco and W. P. Hutzler, Extension of the Gauvin-Tolle optimal value differential stability results to general mathematical programs, Technical Paper T-393, Institute for Management Science and Engineering, The George Washington University, March 1979.

3. J. Gauvin, A necessary and sufficient regularity condition to have bounded multipliers in nonconvex programming, *Math. Programming 12* (1977), 136-138.

4. J. Gauvin, The generalised gradient of a marginal function in mathematical programming, *Math. of Operations Res. 4*(1979), 458-463.

5. J. Gauvin and F. Dubeau, Differential properties of the marginal function in mathematical programming, Technical Report, Ecole Polytechnique de Montréal, June 1979.

6. S. M. Robinson, Stability theory for systems of inequalities, Part II: Differentiable nonlinear systems, *SIAM J. Numer. Anal. 13*(1976), 497-513.

7. R. T. Rockafellar, *Convex Analysis*, Princeton: Princeton University Press, 1970.

Chapter 5 FIRST ORDER SENSITIVITY OF THE OPTIMAL VALUE FUNCTION IN MATHEMATICAL PROGRAMMING AND OPTIMAL CONTROL

HELMUT MAURER / Westfalische Wilhelms-Universität, Münster, Federal Republic of Germany

ABSTRACT

This paper surveys sensitivity properties of the optimal value function for nonlinear programming problems with differentiable perturbations in the objective function and in the constraints. Upper and lower bounds for the directional derivatives of the optimal value function as well as necessary and sufficient conditions for the existence of the directional derivatives are given. The results are applied to perturbed optimal control problems with state constraints. The theory is illustrated by numerical results for the time optimal control of a nuclear reactor.

I. INTRODUCTION

The purpose of this paper is to survey some recent sensitivity results for the optimal value function in nonconvex mathematical programming. Emphasis is placed on *infinite-dimensional* problems since these include optimal control problems as particular cases.

Consider the following mathematical program (P_a) depending on a perturbation parameter a in a Banach space A:

Minimize $F(x,a)$ (P_a)

subject to $x \in C$ and $G(x,a) \in K$.

Here $F: X \times A \to \mathbb{R}$ and $G: X \times A \to Y$ are Fréchet-differentiable mappings, X and Y are Banach spaces, $C \subset X$ and $K \subset Y$ are nonempty closed convex sets.

The set of feasible points for (P_a) is $S(a) = \{x \in C \mid G(x,a) \in K\}$. The optimal value function $v: A \to \bar{\mathbb{R}}$ is then defined by $v(a) = \inf\{F(x,a) \mid x \in S(a)\}$. Let $\bar{a} \in A$ be a fixed reference parameter and consider for a direction $d \in A$ the directional derivatives:

$$v'(\bar{a};d) = \lim_{t \to 0^+} \frac{1}{t} (v(\bar{a} + td) - v(\bar{a}))$$

$$\overset{+}{v}'(\bar{a};d) = \limsup_{t \to 0^+} \frac{1}{t} (v(\bar{a} + td) - v(\bar{a}))$$

$$\underset{+}{v}'(\bar{a};d) = \liminf_{t \to 0^+} \frac{1}{t} (v(\bar{a} + td) - v(\bar{a}))$$

Our aim is to study the first order sensitivity of v at $\bar{a}$, by which we mean the following problem: use the *necessary* optimality conditions for $(P_{\bar{a}})$ and (1) determine upper and lower bounds for the directional derivatives; (2) find necessary and sufficient conditions for the existence of $v'(\bar{a};d)$.

In recent years, this problem has attracted many researchers in mathematical programming, while it has received less attention in optimal control. We briefly review some work done in this area and refer to these papers for additional references.

Mathematical Programming Gauvin/Tolle [4] and Gauvin [5] have considered finite-dimensional problems with right-hand side perturbations. They obtain upper and lower bounds and derive a sufficient condition for the existence of $v'(\bar{a};d)$. These results have been generalized to arbitrary perturbations in Fiacco/Hutzler [3], Gauvin/Dubeau [6]. Levitin [18,19,20] considers infinite-dimensional problems with constraints consisting of operator equality constraints and finitely many inequality constraints. He obtains upper bounds as well as a necessary and sufficient condition for the existence of $v'(\bar{a};d)$. For infinite-dimensional problems of the type (P_a), Gollan [7,8,9], Maurer [22,23,24] have derived an upper bound, while Lempio/Maurer [16,17] treat both problems (1) and (2) listed above.

The proof methodology in [3-6, 16-20] has roughly the same structure: one linearizes problem $(P_{\bar{a}})$, applies duality theory, and uses appropriate implicit function theorems or stability results for the interplay between the linearized problem and the nonlinear problem. The infinite-dimensional approach in [16,17] has been made possible by the rather general stability results in Alt [1] and Robinson [28,29,30].

Optimal Control One well known aspect of the dynamic programming approach and the geometric approach to optimal control is that the adjoint variable appears as the derivative of the optimal value function with respect to the current state; compare, e.g., Bryson/Ho [2], Leitmann [15]. However, no effort has been spent on finding conditions such that the derivative of the optimal value function actually exists. Peterson [27] treats more general perturbations in optimal control problems with state-control constraints. He obtains formulas for the derivative of the optimal value function but bypasses the existence problem by imposing the rather strong assumption that the optimal solution itself is differentiable with respect to perturbations. Sensitivity results in the spirit of problems (1) and (2) for optimal control problems with state constraints have been derived in Maurer [25] by applying the results in [17]. Upper bounds can also be found in Gollan [8].

In the next two sections we summarize the sensitivity results in [17, 25] for mathematical programming problems and optimal control problems, respectively, and add some speculations on possible extension of these results. Section IV contains a numerical example of an optimal control problem illustrating the theory.

Some of the sensitivity results presented here may be extended to mathematical programming problems with *nondifferentiable* data. A survey of these results, however, is beyond the scope of this paper. The reader is referred to Auslender [32] for finite-dimensional problems and to Aubin/Clarke [33] for infinite-dimensional problems with linear constraints.

II. SENSITIVITY IN MATHEMATICAL PROGRAMMING

The Lagrangian function associated with problem (P_a) is defined by

$$L(x,a,\ell) = F(x,a) + \ell(G(x,a)) \qquad \ell \in Y' \tag{1}$$

where Y' denotes the topological dual of Y. Let $\bar{a} \in A$ be a fixed parameter and consider $(P_{\bar{a}})$ as the *unperturbed* problem. The set of optimal solutions of $(P_{\bar{a}})$ is

$$M = \{\bar{x} \in S(\bar{a}) \mid F(\bar{x},\bar{a}) = v(\bar{a})\}$$

which is assumed to be nonempty. The following assumption is supposed to hold throughout this paper: F is differentiable and G is continuously differentiable in the sense of Fréchet at $(\bar{x},\bar{a})$ for all optimal points $\bar{x} \in M$.

Partial derivatives with respect to x and a are denoted by subscripts. A point $\bar{x} \in M$ is called *regular* if

$$0 \in \operatorname{int}\{G(\bar{x},\bar{a}) + G_x(\bar{x},\bar{a})(C - \bar{x}) - K\} \tag{2}$$

where int denotes the topological interior; M is called regular if every point $\bar{x} \in M$ is regular. If $\bar{x} \in M$ is regular then there exists a *Lagrange multiplier* $\ell \in Y'$ satisfying the Kuhn-Tucker condition (cf. [17,31])

$$\begin{aligned} &L_x(\bar{x},\bar{a},\ell)(x - \bar{x}) \geqq 0 \qquad \text{for all } x \in C \\ &\ell(y - G(\bar{x},\bar{a}) \leqq 0 \qquad \text{for all } y \in K \end{aligned} \tag{3}$$

In case K is a closed convex cone with vertex at the origin, the second condition in (3) is equivalent to

$$\ell(G(\bar{x},\bar{a})) = 0, \ \ell(y) \leqq 0 \qquad \text{for all } y \in K$$

By $\Lambda(\bar{x})$ we denote the set of all Lagrange multipliers associated with $\bar{x}$; $\Lambda(\bar{x})$ is convex and weakly compact and hence bounded in the norm-topology; cf. [17] and Theorem 4.3 in [31].

A. Upper Bounds and Strong Stability

The close relations between Lagrange multipliers and the sensitivity of v at $\bar{a}$ are disclosed in the following theorem.

THEOREM 1 Let $\bar{x}$ be regular and let $d \in A$ be fixed. Then for every $\varepsilon > 0$ there exist $\delta > 0$, $c \geqq 0$, and a curve $x: [0,\delta] \to X$ such that for all $t \in [0,\delta]$

$$\begin{aligned} &x(t) \in S(\bar{a} + td), \ ||x(t) - \bar{x}|| \leqq ct \\ &v(\bar{a} + td) \leqq F(x(t), \bar{a} + td)) \\ &\qquad \leqq v(\bar{a}) + t \max_{\ell\in\Lambda(\bar{x})} L_a(\bar{x},\bar{a},\ell)d + t\varepsilon \end{aligned} \tag{4}$$

COROLLARY 1 (*i*) If $\bar{x}$ is regular then

$$\overset{+}{v}'(a;d) \leqq \max_{\ell\in\Lambda(\bar{x})} L_a(\bar{x},\bar{a},\ell)d \tag{5}$$

(*ii*) If M is regular then

$$\overset{+}{v}'(a;d) \leqq \inf_{\bar{x}\in M} \max_{\ell\in\Lambda(\bar{x})} L_a(\bar{x},\bar{a},\ell)d \tag{6}$$

provided that the expression on the right-hand side is finite.

The estimates (5) and (6) have also been obtained by Gollan [7,8,9], Maurer [22,24] with a method different from that in [17]. Gollan has relaxed the regularity assumptions on $\bar{x}$ resp. on M and has derived estimates similar to (5) and (6) in so-called *abnormal* cases. Furthermore, for finite-dimensional problems, he has shown that the set $\Lambda(\bar{x})$ of multipliers ℓ satisfying the *first order* necessary conditions (3) can be replaced by the (usually smaller) set of multipliers satisfying *higher order* necessary conditions. This leads to sharper bounds for $\overset{+}{v}'(a;d)$. It seems worthwhile to generalize these results to infinite-dimensional problems.

We next turn to the problem of finding conditions such that the directional derivative $v'(\bar{a};d)$ exists and, moreover, equality holds in (5) and (6). For that purpose it suffices to give conditions such that for every $\varepsilon > 0$ there exists $\delta > 0$ with

$$v(\bar{a} + td) \geqq v(\bar{a}) + t \max_{\ell\varepsilon\Lambda(\bar{x})} L_a(\bar{x},\bar{a},\ell)d - t\varepsilon$$

for all $t \in [0,\delta]$. Then by reversing inequality (4), there necessarily exists $\delta' \in (0,\delta]$, $c \geqq 0$, and a curve x: $[0,\delta'] \to X$ with $x(t) \in S(\bar{a} + td)$, $||x(t) - \bar{x}|| \leqq ct$, $F(x(t),\bar{a} + td)) \leqq v(\bar{a} + td) + t2\varepsilon$. This motivates the following condition.

Strong Stability Condition Let $d \in A$ be fixed. The strong stability condition holds, if for every $\varepsilon > 0$ there exist $\bar{x} \in M$, $\delta > 0$, $c \geqq 0$, and a curve x: $[0,\delta] \to X$ with

$$\begin{aligned} &x(t) \in S(\bar{a} + td), \ ||x(t) - \bar{x}|| \leqq ct \\ &F(x(t), \bar{a} + td)) \leqq v(\bar{a} + td) + t\varepsilon \end{aligned} \tag{7}$$

for all $t \in [0,\delta]$. The strong stability condition holds at $\bar{x} \in M$ if $\bar{x}$ can be chosen above independently of $\varepsilon > 0$.

This stability condition differs slightly from the one given in Levitin [20]. It is called the *strong* stability condition in order to distinguish it from the *weak* stability condition introduced below. In particular, the strong stability condition (7) holds at $\bar{x}$, if there exists an optimal solution $x(t)$ of $(P_{\bar{a}+td})$ for $t \in [0,\delta]$, $x(0) = \bar{x}$, which is *Lipschitz continuous* at $t = 0$.

THEOREM 2 Let $d \in A$ be fixed. (*i*) Let $\bar{x}$ be regular. Then $v'(\bar{a};d)$ exists and has the representation

$$v'(\bar{a};d) = \max_{\ell \in \Lambda(\bar{x})} L_a(\bar{x},\bar{a},\ell)d \tag{8}$$

if and only if the strong stability condition holds at $\bar{x}$.

(*ii*) Let M be regular and let

$$\inf_{\bar{x} \in M} \max_{\ell \in \Lambda(\bar{x})} L_a(\bar{x},\bar{a},\ell)d \in \mathbb{R}$$

Then $v'(\bar{a};d)$ exists and has the representation

$$v'(\bar{a};d) = \inf_{\bar{x} \in M} \max_{\ell \in \Lambda(\bar{x})} L_a(\bar{x},\bar{a},\ell)d \tag{9}$$

if and only if the strong stability condition holds.

Note that $v'(\bar{a};d)$ could exist without having equality in (8) and (9). Then necessarily the regularity assumption or the strong stability condition is violated; cf. the example given by Gauvin/Tolle (Example 3.1 in [4]), where the strong stability condition does not hold. A similar example is discussed by Gollan [9], who shows in addition that one obtains a sharp bound for $v'(\bar{a};d)$ by using *higher order* necessary conditions. It appears to be an interesting open problem to derive a modified strong stability condition such that (8), (9) hold with $\Lambda(\bar{x})$ replaced by the set of multipliers satisfying higher order necessary conditions. Analyzing examples in this context suggests that such a stability condition should have the form (7), with $||x(t) - \bar{x}|| \leqq ct$ replaced by $||x(t) - \bar{x}|| \leqq ct^{1/k}$, k = order of necessary conditions.

The next theorem contains conditions for the existence of the Gateaux derivative $v'(\bar{a})$, which follows immediately from Theorem 2.

COROLLARY 2 Let $\bar{x}$ be regular. Assume that the strong stability condition holds at $\bar{x}$ and that the set $\{L_a(\bar{x},\bar{a},\ell) \mid \ell \in \Lambda(\bar{x})\} \subset A'$ is a singleton. Then the Gateaux derivative $v'(\bar{a})$ exists and has the representation $v'(\bar{a}) = L_a(\bar{x},\bar{a},\ell)$ for some $\ell \in \Lambda(\bar{x})$.

In particular, the set $\{L_a(\bar{x},\bar{a},\ell) \mid \ell \in \Lambda(\bar{x})\}$ is a singleton if the set $\Lambda(\bar{x})$ is a singleton. This follows by imposing the regularity condition that $G_x(\bar{x},\bar{a}): X \to Y$ is surjective, which is a strengthening of (2).

B. Lower Bounds and Weak Stability

Lower bounds for $\underset{+}{v}'(\bar{a};d)$ are derived in [3,4,5,6] under the assumption that the feasible sets S(a) are uniformly compact near $\bar{a}$. This is not a realistic assumption in *infinite-dimensional* problems. Instead, we shall work with the following weak stability condition which weakens the strong stability condition (7).

Weak Stability Condition Let $d \in A$ be fixed. The weak stability condition holds, if for every $\varepsilon > 0$ there exist $\bar{x} \in M$, $\delta > 0$, and a curve $x: [0,\delta] \to X$ with

$$x(t) \in S(\bar{a} + td) \qquad \lim_{t \to 0^+} x(t) = x(0) = \bar{x}$$
$$F(x(t), \bar{a} + td) \leqq v(\bar{a} + td) + t\varepsilon \tag{10}$$

for all $t \in [0,\delta]$. The weak stability condition holds at $\bar{x} \in M$ if $\bar{x}$ can be chosen above independently of $\varepsilon > 0$.

We note that the weak stability condition holds at $\bar{x}$ if there exists an optimal solution x(t) of $(P_{\bar{a}+td})$ for $t \in [0,\delta]$, $x(0) = \bar{x}$, which is *continuous* at $t = 0$.

THEOREM 3 Let $d \in A$ be fixed. (*i*) Let $\bar{x}$ be regular. Assume that F is continuously differentiable at $(\bar{x},\bar{a})$ and that the weak stability condition holds at $\bar{x}$. Then

$$\min_{\ell \in \Lambda(\bar{x})} L_a(\bar{x},\bar{a},\ell)d \leqq \underset{+}{v}'(\bar{a};d) \tag{11}$$

(*ii*) Let M be regular. Assume that F is continuously differentiable at $(\bar{x},\bar{a})$ for all $\bar{x} \in M$ and that the weak stability condition holds. Then

$$\inf_{\bar{x} \in M} \min_{\ell \in \Lambda(\bar{x})} L_a(\bar{x},\bar{a},\ell)d \leqq \underset{+}{v}'(a;d) \tag{12}$$

Corollary 1 and Theorem 3 can be combined to give a criterion for the existence of the directional derivative $v'(\bar{a};d)$ which is different from the one in Theorem 2.

COROLLARY 3 Let $d \in A$ be fixed. (*i*) Under the assumption of Theorem 3(*i*), suppose further that $\Lambda(\bar{x})$ consists of a single Lagrange multiplier. Then $v'(\bar{a};d)$ exists and has the representation

$$v'(\bar{a};d) = L_a(\bar{x},\bar{a},\ell)d$$

(*ii*) Under the assumption of Theorem 3(*ii*), suppose further that $\Lambda(\bar{x})$ consists of a single Lagrange multiplier $\ell(\bar{x})$ for all $\bar{x} \in M$ and that

$$\inf_{\bar{x}\in M} L_a(\bar{x},\bar{a},\ell(\bar{x}))d \in \mathbb{R}$$

Then $v'(\bar{a};d)$ exists and has the representation

$$v'(\bar{a};d) = \inf_{\bar{x}\in M} L_a(\bar{x},\bar{a},\ell(\bar{x}))d$$

C. Extensions

In finite-dimensional problems Gauvin [5], Gauvin/Dubeau [6] have used the estimates (6), (12) to show that v is locally Lipschitz near $\bar{a}$ and that the generalized gradient $\partial v(\bar{a})$ of v at $\bar{a}$ satisfies

$$\partial v(\bar{a}) \subset \mathrm{co}\left(\bigcup_{\bar{x}\in M} \bigcup_{\ell\in\Lambda(\bar{x})} L_a(\bar{x},\bar{a},\ell)\right) \tag{13}$$

where co stands for convex hull.

Based on Corollary 1 and Theorem 3 it is reasonable to expect similar results for problems with spaces X,Y of arbitrary dimension and a finite-dimensional perturbation space A. The finite dimension of A is needed in constructing a Lipschitz constant for v near $\bar{a}$ with the aid of (6), (12). We are currently studying this problem in greater detail.

III. SENSITIVITY IN OPTIMAL CONTROL

Consider the following control problem with state constraints which depends on a parameter a in a Banach space A:

(OC_a) Determine a control function $u \in L_\infty^m[0,T]$ which minimizes the functional

$$F(x,u,a) = \int_0^T f_0(x,u,a)dt \tag{14}$$

subject to the constraints

$$\dot{x} = f(x,u,a), \quad \psi_0(x(0),a) = 0, \quad \psi_1(x(T),a) = 0 \tag{15}$$

$$u(t) \in U \qquad \text{for a.e. } t \in [0,T] \tag{16}$$

$$S(x,a) \leqq 0 \tag{17}$$

The functions $f_0: \mathbb{R}^{n+m} \times A \to \mathbb{R}$, $f: \mathbb{R}^{n+m} \times A \to \mathbb{R}^n$, $\psi_i: \mathbb{R}^n \times A \to \mathbb{R}^{k_i}$ ($k_i \leqq n$, $i = 0,1$), and $S: \mathbb{R}^n \times A \to \mathbb{R}^s$ are assumed to be continuously differentiable. The admissible control set $U \subset \mathbb{R}^m$ is a closed convex set with nonempty topological interior int U. The end time $T > 0$ is fixed; $L_\infty^m[0,T]$ denotes the space of $\mathbb{R}^m$-valued measurable, essentially bounded, functions defined on $[0,T]$. Many other types of control problems may be reduced to the above form by introducing additional state variables.

The control problem (OC_a) is amenable to the theory developed in the preceding section by the performance of two steps: first, one transforms (OC_a) into an optimization problem (P_a) and, secondly, one computes the partial derivative L_a of the Lagrangian. The second step requires an explicit representation of the Lagrange multiplier ℓ, which is given below.

The literature contains several slightly different methods for treating (OC_a) as an optimization problem (P_a); compare, e.g., [10,12,13]. In order to obtain a convenient representation of the multiplier ℓ we extend the approach of [14]. Let

$$X = W_{1,\infty}^n[0,T] \times L_\infty^m[0,T]$$

$$Y = L_\infty^n[0,T] \times \mathbb{R}^{k_0} \times \mathbb{R}^{k_1} \times C^s[0,T]$$

$$C = \{(x,u) \in X \mid u(t) \in U \text{ for a.e. } t \in [0,T]\}$$

$$K = \{0\} \times C^s[0,T]_- \subset Y$$

where $W_{1,\infty}^n[0,T] = \{x\colon [0,T] \to \mathbb{R}^n \text{ absolutely continuous}, \dot{x} \in L_\infty^n[0,T]\}$; X is the space when endowed with the L_∞-norm

$$||(x,u)||_\infty := \max\{||x||_\infty, ||\dot{x}||_\infty, ||u||_\infty\}$$

The closed convex cone of functions which are nonpositive on $[0,T]$ is denoted by $C^s[0,T]_-$. Define the function $F\colon X \times A \to \mathbb{R}$ as in (14), and the function $G\colon X \times A \to Y$ in view of (15), (17) by

$$G(x,u,a) = \big(\dot{x}(\cdot) - f(x(\cdot),u(\cdot),a),\ \psi_0(x(0),a),\ \psi_1(x(T),a),\ S(x(\cdot),a)\big) \tag{18}$$

Then F and G are continuously differentiable. With these definitions, (OC_a) has been turned into a particular problem (P_a).

In the sequel, all vectors are column vectors if not otherwise stated. The Jacobian matrices of partial derivatives with respect to x and u are denoted by subscripts. Observe that, e.g., f_{0x} is a *row* vector. The transpose of vectors or matrices is denoted by an asterisk. Let $(\bar{x},\bar{u})$ be an optimal solution of $(OC_{\bar{a}})$. Arguments of functions involving $\bar{x}(t)$, $\bar{u}(t)$, and $\bar{a}$ will henceforth be abbreviated by $[t]$. A sufficient condition for the regularity of $(\bar{x},\bar{u})$, i.e., for condition (2), is the following [25].

Regularity Condition 1

(a) The linearized system $\dot{x} = f_x[t]_x + f_u[t]u$ is completely controllable and $\psi_{0x}[0]$, $\psi_{1x}[T]$ have maximal rank;
(b) There exist $(x,u) \in X$ and $\varepsilon > 0$ such that (i) $\{u \in \mathbb{R}^m \mid ||u - u(t)|| < \varepsilon\} \subset U - \bar{u}(t)$ for a.e. $t \in [0,T]$, (ii) $\dot{x} = f_x[t]x + f_u[t]u$, $\psi_{0x}[0]x(0) = 0$ and $S[t] + S_x[t]x(t) < 0$ for $t \in [0,T]$.

A. Representation of Lagrange Multipliers

The Hamiltonian function for problem (OC_a) is

$$H(x,u,a) = f_0(x,u,a) + \lambda^* f(x,u,a) \qquad \lambda \in \mathbb{R}^n \tag{19}$$

The Kuhn-Tucker condition (3) applied to $(OC_{\bar{a}})$ yields a Lagrange multiplier $\ell = (\ell_1,\sigma_0,\sigma_1,\ell_2) \in L_\infty^n[0,T]' \times \mathbb{R}^{k_0} \times \mathbb{R}^{k_1} \times C^s[0,T]'$, where the linear functionals ℓ_1 and ℓ_2 have the following representation:

$$\ell_2(y) = \int_0^T d\mu(t)^* y(t) \qquad \text{for } y \in C^s[0,T] \tag{20}$$

where $\mu\colon [0,T] \to \mathbb{R}^s$ is a function of bounded variation normalized by $\mu(0) = 0$. Moreover, in view of (3), μ is nondecreasing and every component μ_i is constant on intervals with $S_i[t] < 0$, $i = 1,\ldots,s$;

$$\ell_1(y) = -\int_0^T \lambda(t)^* y(t)dt \qquad \text{for } y \in L_\infty^n[0,T] \tag{21}$$

where the adjoint function $\lambda\colon [0,T] \to \mathbb{R}^n$ is the solution of the integral equation

$$\begin{gathered}\lambda(t_1) - \lambda(t_0) = -\int_0^T \left\{H_x[t]^* dt + S_x[t]^* d\mu(t)\right\} \\ \text{for all } t_0 < t_1 \text{ in } [0,T] \\ \lambda(0) = -\psi_{0x}[0]^*\sigma_0, \quad \lambda(T) = \psi_{1x}[T]^*\sigma_1\end{gathered} \tag{22}$$

Furthermore, the *minimal principle* holds:

$$\min_{u \in U} H(\bar{x}(t),u,\bar{a},\lambda(t)) = H[t] = \text{const.} \qquad \text{for a.e. } t \in [0,T] \tag{23}$$

Under additional assumptions on the state constraint $S(x,\bar{a}) \leqq 0$ one can show that the function μ in (22) is differentiable on the boundary of the state constraint $(S(\bar{x}(t),\bar{a}) = 0)$. We briefly summarize such conditions in the case that u and S are scalar, i.e., m = s = 1 in (15)-(17), and that $f(\cdot,\cdot,\bar{a})$ and $S(\cdot,\bar{a})$ are C^∞-functions; the reader is referred to [13,22] for more detail on the subject.

Let S^i $(i \geqq 0)$ denote the ith time derivative of $S(x(t),\bar{a})$ along a trajectory (x,u) of (15). Then as a function of x and u, $S^i = S^i(x,\bar{a})$ $(i = 0,\ldots,p-1)$, $S^p = S^p(x,u,\bar{a})$, where p is the *order of the state constraint* $S(x,\bar{a})$; cf. [13]. Assume that the active set I = $\{t \in [0,T] \mid S(\bar{x}(t),\bar{a}) = 0\}$ consists of finitely many boundary intervals and contact points with the boundary. Finally, suppose that on a boundary interval $[t_1,t_2]$,

$$\begin{aligned} (S^p)_u[t] &\neq 0 \qquad \text{for } t \in [t_1,t_2] \\ \bar{u}(t) &\in \text{int } U \qquad \text{for } t \in (t_1,t_2) \end{aligned} \tag{24}$$

Then μ is a C^∞-function on (t_1,t_2) and $\eta := \dot{\mu} \geqq 0$. Define the augmented Hamiltonian by

$$\tilde{H}(x,u,a,\lambda,\eta) = f_0(x,u,a) + \lambda^* f(x,u,a) + \eta S(x,a) \tag{25}$$

for $\lambda \in \mathbb{R}^n$, $\eta \in \mathbb{R}$. Then the adjoint integral equation (22) reduces to the adjoint differential equation

$$\dot{\lambda} = -\tilde{H}_x[t]^* \tag{26}$$

where λ satisfies the following jump conditions at a junction point or contact point t_1 with the boundary:

$$\lambda(t_1^+) = \lambda(t_1^-) - \nu(t_1)S_x[t_1]^*, \quad \nu(t_1) := \mu(t_1^+) - \mu(t_1^-) \geqq 0 \tag{27}$$

B. Computation of the Partial Derivative L_a

Now we are in a position to compute the expression $L_a(\bar{x},\bar{u},\bar{a},\ell)$ for an optimal solution $(\bar{x},\bar{u})$ of $(OC_{\bar{a}})$. Using the definitions (1), (18) and the representations (20), (21) we obtain

$$\begin{aligned} L_a(\bar{x},\bar{u},\bar{a},\ell) &= \sigma_0^* \psi_{0a}[0] + \sigma_1^* \psi_{1a}[T] \\ &\quad + \int_0^T \left\{ H_a[t]dt + d\mu(t)^* S_a[t] \right\} \end{aligned} \tag{28}$$

For state constraints of order p this gives

$$L_a(\bar{x},\bar{u},\bar{a},\ell) = \sigma_0^* \psi_{0a}[0] + \sigma_1^* \psi_{1a}[T] + \int_0^T \tilde{H}_a[t]dt + \sum_i \nu(t_i)S_a[t_i] \tag{29}$$

where $\tilde{H}$ is the Hamiltonian (25), $\nu(t_i) = \mu(t_i^+) - \mu(t_i^-) \geqq 0$, and the summation is extended over the finitely many junction points and contact points t_i.

The last two formulas, in conjunction with the results of Section II, explain the marginal interpretation of the multipliers λ, μ, σ_0, σ_1 as directional derivatives of the optimal value function. Equation (28) contains a number of well known formulas; compare, e.g., [2, (4.2.14)], [15, (1.35), (4.20)], [21, p. 234]. For instance, if $S[0] < 0$ we have $\lambda(0) = (\partial v/\partial x)(\bar{x}(0))^*$, which follows from Corollary 2 and (28) by setting $\psi_0(x(0),a) = x(0) - a$ and observing $\lambda(0) = -\sigma_0$. A relation similar to (28) has also been derived by Peterson [27] in the presence of state-control constraints instead of state constraints. However, the assumptions used in [27] are much stronger than those of Section II.

Finally, we shall specialize (28), (29) in the case where the perturbation parameter a only occurs in a simple state constraint of the form

$$S(x) \leqq a \qquad a \in \mathbb{R} \tag{30}$$

with $S\colon \mathbb{R}^n \to \mathbb{R}$. Then (28), (29) yield

$$L_a(\bar{x},\bar{u},\bar{a},\ell) = -\left(\int_0^T \eta(t)dt + \sum_i \nu(t_i)\right) = -\mu(T) \tag{31}$$

where $\mu(T) \geqq 0$ is the total variation of μ since μ is nondecreasing. From (30) it is clear that the optimal value function v is *nonincreasing*. Hence v is differentiable a.e. on $v^{-1}(\mathbb{R})$. An example with v being *not* differentiable at countable many parameters a is worked on in [25, Section 5].

C. Extensions

The verification of the strong or weak stability condition (7) resp. (10) in the L_∞-norm may become tedious. It is easier to check (7), (10) in the weaker L_1-norm in which, however, the functions F and G are usually not differentiable. We hope that it is possible to obtain a two-norm version of Theorems 2 and 3.

Gollan [8] has improved the upper bounds (5), (6) by using higher order necessary conditions for control problems without the state constraint (17).

IV. A NUMERICAL EXAMPLE: TIME OPTIMAL CONTROL OF A NUCLEAR REACTOR

The following model of the time optimal control of a nuclear reactor is taken from Hassan/Ghonaimy/Malek [11]. This model has been considered in Maurer [26, Example 7.2] mainly from the viewpoint of junction properties of state constraints. Here we shall discuss those aspects of [26] which illustrate the sensitivity results in the preceding sections.

The problem is to minimize the endtime T subject to

$$\begin{array}{lll} \dot{x}_1 = k_1(x_3 - 1)x_1 + k_2x_2 & x_1(0) = n_0 & x_1(T) = n_T \\ \dot{x}_2 = k_1x_1 - k_2x_2 & x_2(0) = n_0k_1/k_2 & x_2(T) = n_Tk_1/k_2 \\ \dot{x}_3 = u & x_3(0) = 0 & x_3(T) = 0 \\ |u(t)| \leqq 0.2 & 0 \leqq t \leqq T & \end{array}$$

and the state constraint of order $p = 2$

$$S(x) = x_1 \leqq a \tag{32}$$

where x_1 is a neutron density, x_2 is delayed neutron concentration, x_3 is reactivity, and the constants are $k_1 = 5.0$, $k_2 = 0.1$, $n_0 = 0.04$, $n_T = 0.06$.

This problem can be transformed to a problem (OC_a) of the type considered in the preceding section by introducing the endtime T as a new state variable.

The optimal *unconstrained* solution $(\bar{x},\bar{u})$, i.e., the optimal solution of the above problem without the state constraint (32), is unique. The optimal control is bang-bang

$$\bar{u}(t) = \begin{cases} 0.2 & t \in [0,t^*) \\ -0.2 & t \in (t^*,t^{**}) \\ 0.2 & t \in (t^{**},T] \end{cases} \tag{33}$$

with $T = 7.047806$, $t^* = 0.4798784 \times T$, $t^{**} = 0.9798784 \times T$. The switching points t^*,t^{**} are the zeros of the *switching* function $\phi(t) = \lambda_3(t)$, which determines the optimal control $\bar{u}$ via the minimum condition (23) as $\bar{u}(t) = 0.2 \text{ sign } (-\phi(t))$. The function $-\phi(t)$ with respect to normalized time t/T is the solid line in Fig. 1. The unconstrained adjoint variable $\bar{\lambda}(t)$ satisfying (22) with $\mu \equiv 0$ resp. (26), (27) with $\eta = \nu(t_1) = 0$ is unique and its initial value is $\bar{\lambda}(0) = -(2.970144, 2.845469, 5.)^*$. Let

$$\bar{a} := \max\{\bar{x}(t) \mid 0 \leqq t \leqq T\} = \bar{x}(t_1) = 0.1213660 \tag{34}$$

where $t_1 = 0.5429932 \times T$ is a contact point with the boundary for the

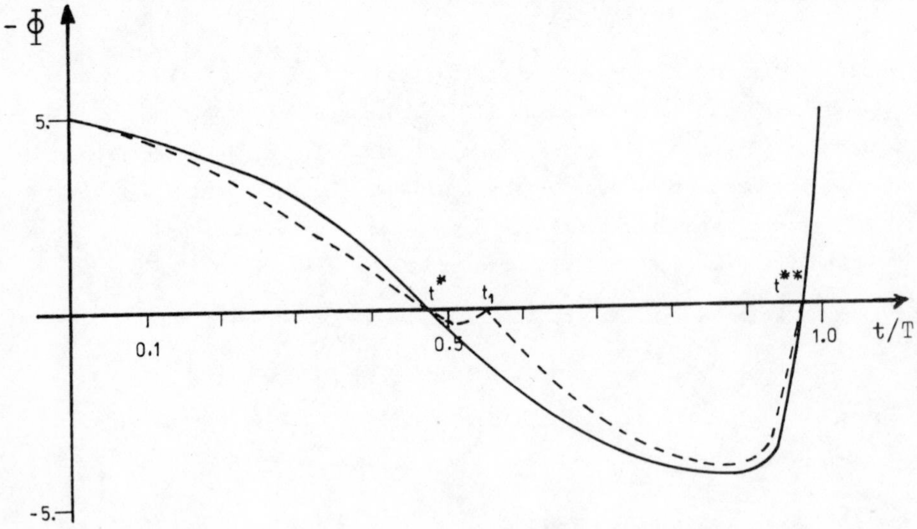

Figure 1 Unconstrained and limiting constrained switching function for $a = \bar{a}$.

trivial state constraint $x_1 \leqq \bar{a}$. Hence $\bar{a}$ is the parameter where the transition from unconstrained to constrained optimal solutions takes place. Numerical computations show that the regularity condition 1 is satisfied at $\bar{a}$.

We shall focus our attention now on studying the behavior of v at $\bar{a}$. Numerical results suggest that v is *not* differentiable at $\bar{a}$ but is differentiable for $a < \bar{a}$. It is clear from the definition of $\bar{a}$ that $v'(\bar{a};1) = 0$. In order to get estimates for $\overset{+}{v}'(\bar{a};-1)$ or the possible left-sided derivative $v'(\bar{a}-) = -v'(\bar{a};-1)$, we have to analyze the adjoint variables $\lambda(t)$ corresponding to the trivial state constraint $x_1 \leqq \bar{a}$. Here the important fact to note is that the adjoint variable $\lambda(t)$ is *not* unique, whereas the unconstrained $\bar{\lambda}(t)$ is unique. This is due to the fact that the jump condition (27) evaluated at the point t_1 in (34) gives

$$\lambda_1(t_1^+) = \lambda_1(t_1^-) - \nu(t_1) \qquad \nu(t_1) \geqq 0 \tag{35}$$

and thus $\lambda(t)$ may have jumps at t_1 while $\bar{\lambda}(t)$ is continuous at t_1.

For a better understanding of the possible jumps $\nu(t_1)$ in (35) we shall describe the behavior of the optimal solution (x_a, u_a) for nontrivial state constraints $x_1 \leqq a$ with $n_T < a < \bar{a}$. The contact point t_1 for $a = \bar{a}$ splits up into a boundary interval $[t_1, t_2]$ for $a < \bar{a}$ and the optimal control u_a is

$$u_a(t) = \begin{cases} 0.2 & t \in [0,t^*) \\ -0.2 & t \in (t^*,t_1) \\ -k_2x_3(t) & t \in [t_1,t_2] \\ -0.2 & t \in (t_2,t^{**}) \\ 0.2 & t \in (t^{**},T] \end{cases} \tag{36}$$

The boundary control $u_a(t) = -k_2x_3(t)$ is obtained from $\ddot{x}_1(t) = 0$ and satisfies (24). The five parameters t^*, t_1, t_2, t^{**}, T are uniquely determined by the three final conditions for $x(T)$ and the two entry conditions $x_1(t_1) = a$, $\dot{x}_1(t_1) = 0$. Furthermore, the entry condition $\phi(t_1) = 0$ must hold [25].

Performing the limit $a \uparrow \bar{a}$, i.e., $t_1 - t_2 \to 0$ in (36), we get the dashed switching function $\phi(t)$ with $\phi(t_1) = 0$ shown in Figure 1. This switching function is admissible for the trivial state constraint $x_1 \leqq \bar{a}$ but is not admissible for the unconstrained problem. The adjoint variable $\tilde{\lambda}(t)$ corresponding to the dashed switching function is completely described by its initial value $\tilde{\lambda}(0) = -(3.376317, 3.234608, 5.)^*$ and its jump $\tilde{\nu}(t_1) = 9.056300$ in (35). Then *all* adjoint variables for $x_1 \leqq \bar{a}$ can be parameterized by the possible jump interval $[0,\tilde{\nu}(t_1)]$ as follows: Given $\nu(t_1) \in [0,\tilde{\nu}(t_1)]$ there exists $\lambda(t)$ with prescribed jump $\nu(t_1)$ in (35). Thus $\lambda(t)$ ranges from the unconstrained $\bar{\lambda}(t)$ with $\nu(t_1) = 0$ and $\phi(t_1) \neq 0$ to the constrained $\tilde{\lambda}(t)$ with jump $\tilde{\nu}_1(t_1)$, for which the entry condition $\phi(t_1) = 0$ holds.

These considerations, together with (5) for $d = -1$ and (31), yield the estimate

$$\overset{+}{v}'(\bar{a};-1) \leqq \max \nu(t_1) = \tilde{\nu}(t_1) = 9.056300$$

while (11) gives the estimate $\underset{+}{v}'(\bar{a};-1) \geqq \min \nu(t_1) = 0$, which is trivial since v is nonincreasing. By Theorem 2 we have $v'(\bar{a}-) = -v'(\bar{a};-1) = -9.056300$ if the strong stability condition (7) holds. This condition cannot be checked analytically, but numerical results indicate that the optimal solution (x_a,u_a) is Lipschitz-continuous at $\bar{a}$ in the L_1-norm, and hence (7) holds in the L_1-norm; cf. the remarks in Section II.C.

Finally, note that the subgradient formula (13) gives

$$\partial v(\bar{a}) \subset [-\tilde{\nu}(t_1),0]$$

REFERENCES

1. W. Alt, Stabilität mengenwertiger abbildungen mit anwendungen auf nichtlineare optimierungsprobleme, Ph.D. Thesis, Mathematisches Institut der Universität Bayreuth, 1979.

2. A. E. Bryson and Y. C. Ho, *Applied Optimal Control*, Waltham, Massachusetts: Blaisdell, 1969.

3. A. V. Fiacco and W. P. Hutzler, Extension of the Gauvin-Tolle optimal value differential stability results to general mathematical programs, Technical Paper T-393, Institute for Management Science and Engineering, The George Washington University, April 1979.

4. J. Gauvin and J. W. Tolle, Differential stability in nonlinear programming, *SIAM J. Control and Optimization 15*(1977), 294-311.

5. J. Gauvin, The method of parametric decomposition in mathematical programming: The nonconvex case, in *Nonsmooth Optimization* (C. Lemarechal and R. Mifflin, eds.), 131-149, Dublin: Pergamon Press, 1978.

6. J. Gauvin and F. Dubeau, Differential properties of the marginal function in mathematical programming, preprint, Ecole Polytechnique de Montréal, June 1979.

7. B. Gollan, Perturbation theory for abstract optimization problems, Preprint No. 47, Mathematisches Institut der Universität Würzburg, April 1979.

8. B. Gollan, Sensitivity results in optimization with applications to optimal control problems, in *Differential Games and Control Theory III* (P.-T. Liu and E. Roxin, eds.), 153-170, Lecture Notes in Pure and Applied Mathematics 44, New York: Marcel Dekker, 1979.

9. B. Gollan, Sensitivity results in optimization theory, in *Proceedings of the III Symposium on Operations Research* (W. Oettli and F. Steffens, eds.), *Operations Res. Verfahren 31*(1979), 247-253.

10. H. Halkin, Necessary conditions for optimal control problems with differentiable or nondifferentiable data, in *Mathematical Control Theory* (W. A. Coppel, ed.), 77-118, Lecture Notes in Mathematics 680, New York: Springer-Verlag, 1978.

11. M. A. Hassan, M. A. R. Ghonaimy, and N. R. Abdel Malek, Computational solution of the nuclear reactor minimum time start-up problem with state-space constraints, in *Proceedings of the 2nd IFAC Symposium on Multivariable Technical Control Systems*, Amsterdam: North-Holland, 1971.

12. A. D. Ioffe and V. M. Tichomirov, *Theory of Extremal Problems*, Amsterdam: North-Holland, 1979.

13. D. H. Jacobson, M. M. Lele, and J. L. Speyer, New necessary conditions of optimality for control problems with state variable inequality constraints, *J. Math. Anal. and Appl. 35*(1971), 255-284.

14. A. Kirsch, W. Warth, and J. Werner, *Notwendige Optimalitätsbedingungen und ihre Anwendung*, Lecture Notes in Economics and Mathematical System 152, New York: Springer-Verlag, 1978.

15. G. Leitmann, *An Introduction to Optimal Control*, San Francisco: McGraw-Hill, 1966.

16. F. Lempio and H. Maurer, Differentiable perturbations of infinite optimization problems, in *Optimization and Operations Research* (R. Henn, B. Korte, and W. Oettli, eds.), 181-191, Lecture Notes in Economics and Mathematical System 157, New York: Springer-Verlag, 1978.

17. F. Lempio and H. Maurer, Differential stability in infinite-dimensional nonlinear programming, *Appl. Math. and Optimization* 6(1980), 139-152.

18. E. S. Levitin, On differential properties of the optimal value of parametric problems of mathematical programming, *Soviet Math. Dokl. 15* (1974), 603-608.

19. E. S. Levitin, On the local perturbation theory of a problem of mathematical programming in a Banach space, *Soviet Math. Dokl. 16*(1975), 1354-1358.

20. E. S. Levitin, Differentiability with respect to a parameter of the optimal value in parametric problems of mathematical programming, *Kibernetika* (1976), 44-59.

21. D. G. Luenberger, *Optimization by Vector Space Methods*, New York: Wiley, 1969.

22. H. Maurer, *Optimale Steuerprozesse mit Zustandsbeschränkungen*, Habilitationsschrift, Mathematisches Institut der Universität Würzburg, 1976.

23. H. Maurer, Contributions to perturbation theory of infinite nonlinear *on Mathematical Programming*, Budapest, 1976, *Survey of Mathematical Programming* (A. Prékopa, ed.), Amsterdam: North Holland, 1980.

24. H. Maurer, A sensitivity result for infinite nonlinear programming problems, I: Theory; II: Applications to optimal control problems, Preprints Nos. 21, 22, Mathematisches Institut der Universität Würzburg, January 1977.

25. H. Maurer, Differential stability in optimal control problems, *Appl. Math. and Optimization* 5(1979), 283-295.

26. H. Maurer, On optimal control problems with bounded state variables and control appearing linearly, *SIAM J. Control and Optimization 15* (1977), 345-362.

27. D. W. Peterson, On sensitivity in optimal control problems, *J. Optimization Theory and Appl. 13*(1974), 56-73.

28. S. M. Robinson, Stability theory for systems of inequalities, Part I: Linear systems, *SIAM J. Numer. Anal. 12*(1975), 754-769.

29. S. M. Robinson, Stability theory for systems of inequalities, Part II: Differentiable nonlinear systems, *SIAM J. Numer. Anal. 13*(1976), 497-513.

30. S. M. Robinson, Regularity and stability for convex multivalued functions, *Math. of Operations Res. 1*(1976), 130-143.

31. J. Zowe and S. Kurcyusz, Regularity and stability for the mathematical programming problem in Banach space, *Appl. Math. and Optimization 5* (1979), 49-62.

32. A. Auslender, Differentiable stability in nonconvex and nondifferentiable programming, *Mathematical Programming Study 10*(1979), 29-41.

33. J. P. Aubin and F. H. Clarke, Shadow prices and duality for a class of optimal control problems, *SIAM J. Control Optimiz. 17*(1979), 567-586.

Chapter 6 REGIONS OF STABILITY FOR ARBITRARILY PERTURBED CONVEX PROGRAMS*

SANJO ZLOBEC / McGill University, Montreal, Quebec, Canada
R. GARDNER / University of Petroleum and Minerals, Dhahran, Saudi Arabia†
ADI BEN-ISRAEL / University of Delaware, Newark, Delaware

ABSTRACT

Using the minimal index set of binding constraints and cones of directions of constancy, we identify three regions of stability for arbitrarily perturbed convex programs. In these regions the optimal solutions and optimal values depend continuously on data.

I. INTRODUCTION

Stability in mathematical programming (i.e., dependence of optimal solutions and values on data) has been extensively studied in the literature, e.g., [1-7]. The case usually treated is the one of the right-hand side perturbations for which there exists a satisfactory theory with economic interpretations (e.g., [7,8,9,10,11]) and remarkable numerical results (e.g., [12]). The case of perturbations of both the right-hand side coefficients and the technological coefficients has been studied for linear

*Research partly supported by the National Research Council of Canada and by NSF Grant ENG77-10126.

†Affiliation at time of writing: Auburn University, Auburn, Alabama.

programs and rather general, though nonconstructive, results have been obtained in [1,2,13]. The literature on questions of stability for convex programs with perturbations in all data is less tractable except where regularity (i.e., Slater's condition) is assumed, e.g., [14].

In this paper we consider arbitrarily perturbed convex programs (P,θ) around a fixed θ^0. Using computable objects from convex programming, such as the minimal index set of binding constraints and cones of directions of constancy (see [15]), we identify and study three regions of stability emanating from θ^0. In these regions the optimal solutions and optimal values always depend continuously on data. Arbitrarily perturbed convex programs possess rather nonintuitive features; some of them are recalled and illustrated in Sec. II. General properties of perturbed (not necessarily convex) programs are reviewed in Sec. III. The results are used there to establish convergence of discretization schemes for semi-infinite convex programs. The three regions of stability are introduced in Sec. IV. There are situations when these regions reduce to a single point (see also [16, Example 2]). If Slater's condition holds for (P,θ^0), the stability regions become a neighborhood of θ^0. In Sec. V the three regions are compared and conditions for their convexity are established. Situations where an optimal solution of (P,θ^0) is not unique are studied in Sec. VI. Tikhonov's regularization [17,18] is modified with the parameter θ running over the regions of stability. A unique minimal-norm type of optimal solution of (P,θ^0) is then obtained as the limit of optimal solutions of a sequence of perturbed programs, each having a unique optimum. Finally, in Sec. VII, we formulate and demonstrate a method for computing two regions of stability.

II. PRELIMINARIES

Consider the *perturbed* program

$$\inf_{(x)} f^0(x,\theta) \qquad\qquad (P,\theta)$$
$$\text{s.t. } f^j(x,\theta) \leqq 0 \qquad j \in \mathcal{P} \triangleq \{1,\ldots,p\}$$

where $x \in R^n$, $\theta \in R^m$ and $\{f^j: j = 0,1,\ldots,p\}$ are continuous in both x,θ. If $\{f^j(\cdot,\theta), j = 0,1,\ldots,p\}$ are convex (resp. linear) functions, $R^n \to R$ for each θ, then (P,θ) is a perturbed *convex* (resp. *linear*) program.

For a given θ we denote:

The *feasible set* $F(\theta) \stackrel{\Delta}{=} \{x \in R^n: f^j(x,\theta) \leqq 0,\ i \in P\}$

The *optimal value* (whenever it exists) $\tilde{f}(\theta)$

The *optimal set* $\tilde{F}(\theta) \stackrel{\Delta}{=} \{x \in F(\theta): f^0(x) = \tilde{f}(\theta)\}$

We will occasionally assume that Slater's condition holds for (P,θ), where θ is fixed, i.e.,

$$\exists x \in F(\theta) \ni f^j(x,\theta) < 0 \qquad \forall\, j \in P$$

The perturbed programs will be studied around a fixed θ^0. Except for the illustrative examples appearing in this section, it is assumed throughout the paper that the set $\tilde{F}(\theta^0)$ is nonempty and bounded.

The study of perturbed convex programs is rather nonintuitive. Some of the difficulties are listed below.

i. There are situations where every perturbation of θ^0 results in jumps in the optimal values and solutions, even if (P,θ) is linear (e.g., [16, Example 2]).

ii. It may happen that $\tilde{F}(\theta^0) \neq \emptyset$ and that Slater's condition holds for (P,θ^0), but that every perturbation θ of θ^0 results in an empty $\tilde{F}(\theta)$, even if (P,θ) is linear. (See Example 1 below.)

iii. For every perturbation θ of θ^0, the set $\tilde{F}(\theta)$ may be nonempty but unbounded when $\theta \to \theta^0$. Again, (P,θ) may be linear and Slater's condition may hold for (P,θ^0). (See Example 2.)

iv. Convergence $\theta^k \to \theta^0$ does not imply $\tilde{f}(\theta^k) \to \tilde{f}(\theta^0)$, even if (P,θ) is convex and int $F(\theta^0)$ is nonempty [16, Example 3].

v. Uniform convergence of a sequence of constraint functions does not imply convergence of the associated sequence of feasible sets. (See Example 3, also [3, p. 10] for a slightly more complicated program.)

These are illustrated by the following examples.

EXAMPLE 1.

$$\text{Min } -x_1$$
$$\text{s.t. } x_1 + \theta x_2 \leqq 1$$

Here $\tilde{F}(\theta^0) = \{(1,x_2)^T\}$ at $\theta^0 = 0$, but $\tilde{F}(\theta) = \emptyset$ for $\theta \neq \theta^0$.

EXAMPLE 2.

$$\begin{aligned} &\text{Min } -x_1 \\ &\text{s.t. } x_1 + \theta x_2 \leqq 0 \\ &\quad\;\; x_1 \qquad\; \leqq 1 \end{aligned}$$

For every $\theta \neq \theta^0 = 0$, the optimal solution is $x_1 = 1$, $x_2 = -\frac{1}{\theta}$. When $\theta \to \theta^0$, $\tilde{F}(\theta)$ does not have a limit point.

EXAMPLE 3. Consider (P,θ) with the single constraint

$$f^1(x,\theta) = \theta^2 x \leqq 0$$

When $\theta \to \theta^0 = 0$, $f^1(x,\theta) \to 0$ uniformly on any finite interval. But $F(\theta) \not\to F(\theta^0)$.

III. GENERAL PROPERTIES OF PERTURBED PROGRAMS

For a study of stability we need the concept of the convergence of sets. Let $\{E^k\}$ be an infinite sequence of sets in a space X and denote:

$$\overline{\lim_{k\to\infty}} E^k \triangleq \left\{x \in X\colon x = \lim_{i\to\infty} x^{k,i},\ \{k,i\} \text{ is an infinite subsequence of the integers and } x^{k,i} \in E^{k,i}\right\};\text{ and}$$

$$\underline{\lim_{k\to\infty}} E^k \triangleq \left\{x \in X\colon x = \lim_{k\to\infty} x^k, \text{ where } x^k \in E^k \text{ for all but a finite number of } k\right\}.$$

Both $\overline{\lim_{k\to\infty}} E^k$ and $\underline{\lim_{k\to\infty}} E^k$ are closed and $\underline{\lim_{k\to\infty}} E^k \subset \overline{\lim_{k\to\infty}} E^k$.

If $\underline{\lim_{k\to\infty}} E^k = \overline{\lim_{k\to\infty}} E^k \triangleq E$, we say that $E^k \to E$.

Let $\theta^k \to \theta^0$. Our objective is to find conditions under which some or all of the following properties hold:

(A) $\tilde{F}(\theta^k) \to \tilde{F}(\theta^0)$.

(B) $\tilde{F}(\theta^k) \neq 0$ for large k; if $x^k \in \tilde{F}(\theta^k)$ then the sequence $\{x^k\}$ is bounded and all its limit points are in $\tilde{F}(\theta^0)$.

(C) $\tilde{F}(\theta^k) \neq 0$ for large k and $\tilde{f}(\theta^k) \to \tilde{f}(\theta^0)$.

(D) $F(\theta^k) \to F(\theta^0)$.

Note that only (D) is independent of f^0. It is clear that (A) $\Longrightarrow$ (B), by boundedness and nonemptiness of of $\tilde{F}(\theta^0)$, and that (B) $\Longrightarrow$ (C). If f^0 is convex in x, then, in fact, (B) $\Longleftrightarrow$ (C). In order to prove this statement first we need the following lemma.

LEMMA 1

$$\overline{\lim_{k\to\infty}}\, F(\theta^k) \subset F(\theta^0)$$

PROOF. Suppose $x \in \overline{\lim\limits_{k\to\infty}}\, F(\theta^k)$. Then there are $x^i \in F(\theta^{k,i})$ such that $x^i \to x$. So, for each i, $f^j(x^i,\theta^{k,i}) \leqq 0$, $j \in P$. By continuity $f^j(x,\theta^0) \leqq 0$, $j \in P$. Thus $x \in F(\theta^0)$.

THEOREM 1 If $f^0(\cdot,\theta)$ is convex for every θ, then (C) $\Longrightarrow$ (B).

PROOF. Let $x^k \in \tilde{F}(\theta^k)$. Since $\tilde{f}(\theta^k) \to \tilde{f}(\theta^0)$, for a given $\varepsilon < 0$ there is a $\kappa > 0$ such that

$$\tilde{f}(\theta^k) < \tilde{f}(\theta^0) + \varepsilon$$

for $k \geqq \kappa$. As $f^0(x^k) = \tilde{f}(\theta^k)$, and f^0 is convex, the sequence $\{x^k\}$ is bounded, by, e.g., [2, Lemma 26.1]. Suppose that $\bar{x}$ is a limit point of $\{x^k\}$, i.e., $x^{k,i} \to \bar{x}$. Then

$$f^0(x^{k,i}) = \tilde{f}(\theta^{k,i}) \to \tilde{f}(\theta^0)$$

since (C) holds, and

$$f^0(x^{k,i}) \to f^0(\bar{x})$$

by continuity of f^0. So $f^0(\bar{x}) = \tilde{f}(\theta^0)$. Also, by Lemma 1, $\bar{x} \in F(\theta^0)$. Thus $\bar{x} \in \tilde{F}(\theta^0)$.

Denote by (P)' the condition that (P) holds for *all* continuous f^0.

THEOREM 2

$$(A)' \Longrightarrow (B)' \Longrightarrow (C)' \Longrightarrow (D)$$

PROOF. The implications (A)' $\Longrightarrow$ (B)' and (B)' $\Longrightarrow$ (C)' are clear. (C)' $\Longrightarrow$ (D): It is enough to show that

$$F(\theta^0) \subset \underline{\lim_{k\to\infty}}\, F(\theta^k)$$

in view of Lemma 1. Suppose that this is not so. Choose

$$p \in F(\theta^0) \setminus \underline{\lim}_{k\to\infty} F(\theta^k)$$

and define $f^0(x,\theta) = ||x - p||$. Since $p \in F(\theta^0)$, $\tilde{f}(\theta^0) = 0$. But $p \notin \underline{\lim}_{k\to\infty} F(\theta^k)$, so there is an $\varepsilon > 0$ such that for each ℓ, there is a $k \geq \ell$ with $F(\theta^k) \cap B_\varepsilon(p) = \emptyset$, where $B_\varepsilon(p)$ is the ball of radius ε with the center at p. Thus for arbitrarily large k, $\tilde{f}(\theta^k) \geq \varepsilon$. So $\tilde{f}(\theta^k) \not\to \tilde{f}(\theta^0)$, a contradiction!

REMARK 1. Fiacco (see [10, Corollary 2.2]) has proved that if $F(\theta^k) \to F(\theta^0)$ and $x^k \in \tilde{F}(\theta^k)$ then every limit point of $\{x^k\}$ is in $\tilde{F}(\theta^0)$. He observes that this result can be deduced from [2, Theorem I.2.2]. Note that (D) does not imply $\tilde{F}(\theta^k) \neq 0$ for large k, so (D) $\not\Rightarrow$ (B)'.

REMARK 2. A counterexample to (D) $\Rightarrow$ (C)' is given in [2, p. 529].

Theorem 2 can be strengthened in the convex case. We will make use of the following result.

LEMMA 2 Consider a perturbed convex program (P,θ) and a sequence $\{\theta^k\}$ converging to θ^0. If $F(\theta^k) \to F(\theta^0)$, then $\tilde{F}(\theta^k) \neq \emptyset$ for large k.

PROOF. Let B_N be the ball of radius N with the center at $x = 0$ and such that $\tilde{F}(\theta^0) \subset B_N$. Choose $x^* \in \tilde{F}(\theta^0)$ and $x^k \in F(\theta^k)$ such that $x^k \to x^*$. Suppose that $\tilde{F}(\theta^{k,i}) = \emptyset$ for $i = 1,2,\ldots$. Then $F(\theta^{k,i})$, being closed, is unbounded and for each i there is a direction d^i such that $x^{k,i} + \alpha d^i \in F(\theta^{k,i})$ for *all* $\alpha > 0$ and $f^0(x^{k,i} + \alpha d^i, \theta^{k,i}) < f^0(x^{k,i}, \theta^{k,i})$. Choose α_i so that $y^i = x^{k,i} + \alpha_i d^i \in B_{2N}\setminus B_N$. Then the sequence $\{y^i\}$ is bounded and it has a convergent subsequence $y^{i,j} \to y^* \in F(\theta^0)\setminus\tilde{F}(\theta^0)$. Now

$$f^0(y^{i,y}, \theta^{k,i,j}) \leq f^0(x^{k,i,j}, \theta^{k,i,j})$$

and hence, by continuity,

$$f^0(y^*,\theta^0) \leq f^0(x^*,\theta^0)$$

This and $y^* \in F(\theta^0)$ imply $y^* \in \tilde{F}(\theta^0)$, a contradiction.

Denote by (P)" the condition that (P) holds for all convex f^0.

THEOREM 3 For a perturbed convex program (P,θ):

$$(A)'' \Rightarrow (B)'' \Leftrightarrow (C)'' \Leftrightarrow (D)$$

PROOF. In view of Theorem 2 it is enough to show that (D) ⟹ (B)". Note firstly that $\tilde{F}(\theta_k) \neq \emptyset$ for large k, by Lemma 2. Translating [2, Theorem I.3.3] into our terminology, we see that if U is open and $\tilde{F}(\theta^0) \subset U$, then $\tilde{F}(\theta^k) \subset U$ for sufficiently large k. Choosing U bounded, it follows that if $x^k \in \tilde{F}(\theta^k)$, then the sequence $\{x^k\}$ is bounded. Further, each limit point of $\{x^k\}$ lies in $\tilde{F}(\theta^0)$; for if x^* is a limit point with $x^* \notin \tilde{F}(\theta^0)$ we can choose U with $x^* \notin U$ and $\tilde{F}(\theta^0) \subset U$ to obtain a contradiction.

The above results are rather general and, therefore, they can be used in a context different from perturbed programs. This will now be demonstrated for (unperturbed) convex programs with infinitely many constraints. We will prove the convergence of discretization methods for solving such problems.

To this end we consider the (semi-infinite) program

$$\begin{aligned} &\inf f^0(x) \\ &\text{x.t. } f(x,t) \leqq 0 \qquad t \in T \end{aligned} \tag{P,T}$$

where f^0, $f(\cdot,t)$: $R^n \to R$ are continuous convex functions and T is an infinite (usually compact) set in R^p. We also assume that $f(\cdot,\cdot)$ is continuous in both variables x and t. The program may be *discretized* by selecting finite sets $T_k \subset T$, k = 1,2,..., such that $T_k \to T$ as $k \to \infty$, and considering the finitely constrained programs

$$\begin{aligned} &\inf f^0(x) \\ &\text{s.t. } f(x,t) \leqq 0 \qquad t \in T_k \end{aligned} \tag{P,T_k}$$

As in Sec. II, we denote, for a given $T_k \subset T$:

The *feasible set* $F(T_k) \overset{\Delta}{=} \{x \in R^n: f(x,t) \leqq 0,\ t \in T_k\}$

The *optimal value* $\tilde{f}(T_k)$

The *optimal set* $\tilde{F}(T_k) \overset{\Delta}{=} \{x \in F(T_k): f^0(x) = \tilde{f}(T_k)\}$

It is of interest to know conditions under which, for example, $\tilde{f}(T_k) \to \tilde{f}(T)$. Surprisingly, there is no further restriction on the choice of T_k.

PROPOSITION 1 Consider the program (P,T) where it is assumed that F(T) is nonempty and bounded. Then for any choice of $T_k \subset T$ such that $T_k \to T$ we have $\tilde{f}(T_k) \to \tilde{f}(T)$; further, if $x^k \in \tilde{F}(T_k)$, then the sequence $\{x^k\}$ is bounded and all its limit points are in $\tilde{F}(T)$.

PROOF. As $T_k \subset T$, $F(T) \subset F(T_k)$ for each k; consequently

$$F(T) \subset \underline{\lim}_{k\to\infty} F(T_k)$$

The appropriate version of Lemma 1 (proved by the same argument) gives

$$\overline{\lim}_{k\to\infty} F(T_k) \subset F(T)$$

so that in fact $F(T_k) \to F(T)$. The result follows from the corresponding form of Theorem 3 ((D) => (C)" => (B)"); again, no new arguments are needed.

Such results are useful in approximation theory; see [19, p. 96].

IV. REGIONS OF STABILITY FOR PERTURBED CONVEX PROGRAMS

Continuity of optimal solutions and of optimal values of (P,θ) at a given θ^0 depends on how the sequence $\{\theta^k\}$ approaches θ^0. For one "trajectory" $\{\theta^k\}$ all sets $F(\theta^k)$ may be empty, while for another, $\tilde{f}(\theta^k)$ may have a jump at θ^0. The results of the preceding section provide conditions for stability. In particular, they determine, in principle, the set of all "trajectories" along which the continuity is preserved. However, these results are nonconstructive and rather difficult to implement. Our objective is to construct regions of stability using the computable objects from convex programming, such as the cones of directions of constancy [20,21] and the minimal index sets of binding constraints [15,22].

Here and in the following sections we will study only perturbed *convex* programs (P,θ). Following [15,16,20,22] we introduce, for every θ, the minimal index set of binding constraints as

$$\mathcal{P}^{=}(\theta) = \{j \in \mathcal{P}: x \in F(\theta) \Rightarrow f^j(x,\theta) = 0\}$$

Note that $\mathcal{P}^{=}(\theta) = \emptyset$ if and only if Slater's condition holds. Using $\mathcal{P}^{=}(\theta)$ we define

$$F^{=}(\theta) = \{x \in R^n: f^j(x,\theta) \leqq 0, \ j \in \mathcal{P}^{=}(\theta)\}$$

Note that $F^{=}(\theta) \supset F(\theta)$. Using convexity arguments one can show that

$$F^{=}(\theta) = \{x \in R^n: f^j(x,\theta) = 0,\ i \in \mathcal{P}^{=}(\theta)\}$$

When it is clear from the context that θ is specified, we will omit the argument θ, i.e., we will write $\mathcal{P}^{=}$, $F^{=}$ instead of $\mathcal{P}^{=}(\theta)$, $F^{=}(\theta)$, etc. The cone of directions of constancy of $f(\cdot,\theta)\colon R^n \to R$ at $x \in R^n$ is defined by (see [20])

$$D_f(x) = \{d \in R^n: \exists \bar{\alpha} > 0 \ni f(x + \alpha d) = f(x),\ \forall\ 0 \leqq \alpha \leqq \bar{\alpha}\}$$

For a large class of "faithfully convex" functions, i.e., functions of the form (for a given θ)

$$f(x) = \phi(Ax + b) + a^T x + \alpha$$

where $\phi\colon R^m \to R$ is strictly convex, $a \in R^n$ and $\alpha \in R$, the cone of directions of constancy is

$$D_f = N\begin{pmatrix} A \\ a^T \end{pmatrix}$$

the null space of $\begin{pmatrix} A \\ a^T \end{pmatrix}$ independent of x. Moreover, for such functions one can show that

$$F^{=} = p + D^{=}$$

for an arbitrary $p \in F$. Here we abbreviate

$$D^{=} = \bigcap_{j \in \mathcal{P}^{=}} D_{f^j}(x)$$

Algorithms for calculating $\mathcal{P}^{=}$ and $D^{=}$ are suggested in [22] (see also [15]) and [21], respectively.

In this paper we are looking for a *region of stability* S of (P,θ) at a fixed θ^0, $\theta^0 \in S$ in the sense of the following definition.

DEFINITION 1 Perturbed convex program (P,θ) is *stable in a region* S *at* $\theta^0 \in S$, if for some neighborhood N_0 of θ^0

(i) $\tilde{F}(\theta) \neq \emptyset$, $\forall \theta \in N_0 \cap S$; and
(ii) $\{\theta^k\} \subset N_0 \cap S$, $\theta^k \to \theta^0$ imply $\tilde{f}(\theta^k) \to \tilde{f}(\theta^0)$.

REMARKS. In view of Theorem 3 one can replace the statements (i) and (ii) in Definition 1 by the equivalent single statement (iii)', if $\theta^k \to \theta^0$,

$\{\theta^k\} \subset N_0 \cap S$ and $x^k \in \tilde{F}(\theta^k)$ then the sequence $\{x^k\}$ is bounded and all its limit points are in $\tilde{F}(\theta^0)$.

Consider the following three sets emanating from θ^0:

$$V_0^= = \left\{ \begin{array}{l} F^=(\theta^0) \subset F^=(\theta) \text{ and} \\ \theta \in R^m\text{: } f^j(x,\theta) \leqq 0,\ \forall\, x \in F(\theta^0) \\ \qquad\qquad j \in P^=(\theta^0),\ j \notin P^=(\theta) \end{array} \right\}$$

$$W_0^= = \{\theta \in R^m\text{: } F^=(\theta^0) \subset F^=(\theta) \text{ and } P^=(\theta^0) = P^=(\theta)\}$$

$$V_0 = \{\theta \in R^m\text{: } F(\theta^0) \subset F(\theta)\}$$

The set $V_0^=$ was introduced in [16]. There it has been shown that (P,θ) is indeed stable in $V_0^=$. Since $W_0^= \subset V_0^=$, also the set $W_0^=$ is a region of stability. We now prove that V_0 is a region of stability (for all continuous convex objective functions). The idea is to show that the property (D) holds, which together with boundedness of $\tilde{F}(\theta)$ and convexity implies stability.

THEOREM 4 Consider a perturbed convex program (P,θ). If $\theta^k \to \theta^0$, $\{\theta^k\} \subset V_0$, then $\tilde{F}(\theta^k) \neq \emptyset$ for large k and $\tilde{f}(\theta^k) \to \tilde{f}(\theta^0)$.

PROOF. Since $F(\theta^0) \subset F(\theta^k)$ and $\theta^k \to \theta^0$, it follows that

$$F(\theta^0) \subset \underline{\lim}_{k\to\infty} F(\theta^k)$$

Hence $F(\theta^k) \to F(\theta^0)$, by Lemma 1. But (D) $\Rightarrow$ (C)", by Theorem 3.

The fact that $W_0^=$ is a region of stability can be proved directly using the next lemma.

LEMMA 3 Consider a perturbed convex program (P,θ) and a sequence $\{\theta^k\}$ converging to θ^0. If $\underline{\lim}\, F^=(\theta^k) \supset F^=(\theta^0)$ and $P^=(\theta^k) \to P^=(\theta^0)$, then $F(\theta^k) \to F(\theta^0)$.

PROOF. Take $p \in \text{rel int } F(\theta^0)$. Then

$$f^j(p,\theta^0) < 0 \qquad \forall\, j \in P\backslash P^=(\theta^0) \tag{1}$$

Since $p \in F^=(\theta^0)$ and $F^=(\theta^0) \subset \underline{\lim}\, F^=(\theta^k)$, there exist $p^k \in F^=(\theta^k)$ with $p^k \to p$. We claim that there is a $\kappa \geqq 0$ such that $p^k \in F(\theta^k)$ for all

$k \geqq \kappa$. If this were not true, there would be a subsequence $\{p^{k,i}\}$ of $\{p^k\}$ such that

$$p^{k,i} \in F^{=}(\theta^{k,i}) \setminus F(\theta^{k,i})$$

i.e.,

$$f^{j_{k,i}}(p^{k,i}, \theta^{k,i}) > 0$$

for some $j_{k,i} \in P \setminus P^{=}(\theta^{k,i})$. Since the minimal index sets of binding constraints are finite, for sufficiently large i, $P^{=}(\theta^{k,i}) = P^{=}(\theta^0)$. Therefore

$$f^{j*}(p^{k,i,\ell}, \theta^{k,i,\ell}) > 0 \qquad \text{for some } j* \in P \setminus P^{=}(\theta^0)$$

and sufficiently large ℓ. This gives in the limit

$$f^{j*}(p,\theta^0) \geqq 0 \qquad j* \in P \setminus P^{=}(\theta^0)$$

which contradicts (1) and the claim is established. Now we use the fact that $F(\theta^0) = \text{cl}(\text{rel int } F(\theta^0))$. Hence for any $p \in F(\theta^0)$, there exist $p^k \in F(\theta^k)$ with $p^k \to p$. So $\underline{\lim}\, F(\theta^k) \supset F(\theta^0)$, which together with Lemma 1 implies $F(\theta^k) \to F(\theta^0)$.

THEOREM 5 For a convex program (P,θ), $\theta^k \to \theta^0$ where $\{\theta^k\} \subset W_0^{=}$ implies that $\tilde{F}(\theta^k) \neq \emptyset$ for large k and $\tilde{f}(\theta^k) \to \tilde{f}(\theta^0)$.

PROOF. Take $\{\theta^k\} \subset W_0^{=}$, $\theta^k \to \theta^0$. Then, by definition, $F^{=}(\theta^k) \supset F^{=}(\theta^0)$, which implies $\underline{\lim}\, F^{=}(\theta^k) \supset F^{=}(\theta^0)$. On the other hand, $P^{=}(\theta^k) = P^{=}(\theta^0)$. Now, by Lemma 3, $F(\theta^k) \to F(\theta^0)$ and the stability follows by Theorem 3.

EXAMPLE 4. This example shows that the assumption on $P^{=}(\theta^k)$ in Lemma 3 cannot be omitted. Consider

$$f^1 = -|\theta|x \leqq 0$$

around $\theta^0 = 0$. Here

$$F(0) = R, \; P^{=}(0) = \{1\}, \; F^{=}(0) = R$$

and, for any $\theta \neq 0$,

$$F(\theta) = [0,\infty), \; P^{=}(\theta) = \emptyset, \; F^{=}(\theta) = R$$

When $\theta^k \to 0$, then $F^{=}(\theta^k) \to F^{=}(0)$. Nevertheless, $F(\theta^k) \not\to F(0)$.

If Slater's condition holds for the convex program (P,θ^0), then the program is stable in a neighborhood of θ^0; this was proved in [14]. Here we deduce this statement from our results as follows: When Slater's condition holds, then $P^=(\theta^0) = \emptyset$, $F^=(\theta^0) = R^n$ and further, by continuity of $f^j(x,\cdot)$, $j \in P$, it follows that $P^=(\theta) = \emptyset$, $F^=(\theta) = R^n$ for all θ's close to θ^0. Thus both regions of stability $V_0^=$ and $W_0^=$ reduce to a neighborhood of θ^0.

EXAMPLE 5. This example is taken from [8, p. 120].

$$\inf -\theta x_1 - 6x_2$$
$$\text{s.t. } \theta x_1 + x_2 - \theta \leqq 0$$
$$x_1 + 2x_2 - 8 \leqq 0$$

It is obvious that Slater's condition holds for every fixed value of θ. Therefore, around any θ there is a neighborhood in which the perturbed program is stable. This statement holds true regardless of the objective function (provided that $\tilde{F}(\theta) \neq \emptyset$ and bounded).

V. PROPERTIES OF THE THREE REGIONS OF STABILITY

When Slater's condition holds, $V_0^= = R^m$ and hence $V_0 \subset V_0^=$. The example below shows that the set V_0 may be larger than $V_0^=$.

EXAMPLE 6. Consider

$$f^1 = -x \leqq 0$$
$$f^2 = -\theta(1 - (\theta/|\theta|))(|x| - x) \leqq 0$$
$$f^3 = \theta(1 + (\theta/|\theta|))(|x| - x) \leqq 0$$

where $0/|0|$ is defined to be 1. For every θ, $F(\theta) = [0,\infty)$ and $P^=(\theta) = \{2,3\}$. But

$$F^=(\theta) = \begin{cases} R & \text{if } \theta = 0 \\ [0,\infty) & \text{if } \theta \neq 0 \end{cases}$$

Thus, at $\theta^0 = 0$, $V_0^= = \{\theta^0\}$ while $V_0 = R$.

The region $W_0^=$ is generally different from $V_0^=$ (and V_0). This is demonstrated by the next example.

EXAMPLE 7. Consider

$$f^1 = -|\theta|x + (|x| - x) \leqq 0$$

$$f^2 = -\theta(1 - (\theta/|\theta|))(|x| - x) \leqq 0$$

$$f^3 = \theta(1 + (\theta/|\theta|))(|x| - x) \leqq 0$$

For each θ, $F(\theta) = F^{=}(\theta) = [0,\infty)$. But

$$P^{=}(\theta) = \begin{cases} \{1,2,3\} & \text{if } \theta = 0 \\ \{2,3\} & \text{otherwise} \end{cases}$$

Take $\theta^0 = 0$. Then $V_0^{=} = R$. However, with $\theta^k = (-1)^k/k$, $k = 1,2,\ldots$, $P^{=}(\theta^k) \not\to P^{=}(\theta^0)$, i.e., $\theta^k \notin W_0^{=}$.

The three regions of stability are generally nonconvex. This will be demonstrated for V_0.

EXAMPLE 8. Take

$$f^1 = (\theta^2 - 1)|x| \leqq 0$$

Here $F(-1) = F(1) = R$ while $F(\theta) = \{0\}$ for $-1 < \theta < 1$. So if $\theta^0 = 1$, $V_0 = \{-1,1\}$.

A condition under which V_0 is convex follows.

THEOREM 6 Consider a perturbed convex program (P,θ). Suppose that there is a neighborhood N_0 of θ^0 such that $N_0 \cap \{\theta: f^j(x,\theta) = 0\}$ is convex for each $x \in F(\theta^0)$ and $j \in P$. Then $V_0 \cap N_0$ is convex.

PROOF. Choose θ^1 and θ^2 in $V_0 \cap N_0$, so that $F(\theta^0) \subset F(\theta^i)$, $i = 1,2$. Consider an $x \in F(\theta^0)$; then $f^j(x,\theta^i) \leqq 0$, $i = 1,2$, $j \in P$. Let $\lambda \in [0,1]$ and $\theta = \lambda\theta^1 + (1 - \lambda)\theta^2$. We claim that $f^j(x,\theta) \leqq 0$, $j \in P$. If this were not true, there would exist at least one index $j_0 \in P$ such that $f^{j_0}(x,\theta) > 0$. Then, by continuity, there exist λ_1,λ_2 with $0 \leqq \lambda_2 < \lambda < \lambda_1 \leqq 1$ such that $f^{j_0}(x,\theta_i) = 0$, where $\theta_i' = \lambda_i\theta' + (1 - \lambda_i)\theta^2$, and $i = 1,2$. This contradicts the assumption that $N_0 \cap \{\theta: f^{j_0}(x,\theta) = 0\}$ is convex. So $f^j(x,\theta) \leqq 0$, whence $F(\theta^0) \subset F(\theta)$. This shows $\theta \in V_0 \cap N_0$, which is then convex.

VI. TIKHONOV'S REGULARIZATION

If the convex programs (P,θ) have a unique optimal solution, then the concept of stability can be strengthened.

DEFINITION 2 The perturbed convex program (P,θ) is *optimal-solution-continuous in a region* S *at* $\theta^0 \in S$, if for some neighborhood N_0 of θ^0

(i) $\tilde{F}(\theta) \neq \emptyset$ and $\tilde{F}(\theta)$ consists of a single vector, $\forall\, \theta \in N_0 \cap S$
(ii) if $\theta^k \to \theta^0$, $\{\theta^k\} \subset N_0 \cap S$ and $x^k \in \tilde{F}(\theta^k)$, then $x^k \to x^0 \in \tilde{F}(\theta^0)$

We show in this section how a particular optimal solution $\tilde{x}$ of (P,θ^0) can be obtained as the limit of a sequence of optimal solutions of optimal-solution-continuous programs (P,θ). The particular optimal solution $\tilde{x}$ will be the solution of

$$\begin{aligned}&\min J(x)\\ &\text{s.t. } x \in \tilde{F}(\theta^0)\end{aligned} \tag{2}$$

where J(x) is a strictly convex function. For the choice $J(x) = ||x||^2$, the optimal solution of (2) is clearly the optimal solution of (P,θ^0) having the smallest norm. This general technique is frequently termed Tikhonov's regularization [14,17,18].

Let J(x) be a strictly convex function such that for some scalar δ_0 the set $\{x: J(x) \leqq \delta_0\}$ is nonempty and bounded. With such a choice of the function, the set $\{x: J(x) \leqq \delta\}$ is either bounded or empty, for any scalar δ.

Consider the perturbed convex program

$$\begin{aligned}&\inf_{(x)} f^0(x,\theta) + \varepsilon J(x)\\ &\text{s.t. } f^j(x,\theta) \leqq 0 \qquad j \in P\end{aligned} \tag{T,θ}$$

where $\varepsilon > 0$. Let S denote any of the three regions of stability V_0, $V_0^=$, or $W_0^=$, at a fixed parameter θ^0.

THEOREM 7 If $\tilde{F}(\theta^0)$ is nonempty then (T,θ) is optimal-solution-continuous in a region S at θ^0 for every $\varepsilon > 0$. Moreover, if $x(\varepsilon)$ denotes the optimal solution of (T,θ^0) for a given $\varepsilon > 0$ and if $\tilde{x}$ denotes the optimal solution of (2), then

$$\lim_{\varepsilon \to 0^+} x(\varepsilon) = \tilde{x}$$

PROOF. By the definition of J(x), the set $\tilde{F}(\theta^0)$ is bounded. Using the same arguments as in [16, Theorem 2], but applied to the sequence $\{\theta^k\} \subset S \cap N_0$ (rather than $\{\theta^k\} \subset V_0^= \cap N_0$), one concludes that (T,θ) is stable in S at θ^0. But (T,θ^0) has a unique minimum, since the function

$$\Phi_\varepsilon(x,\theta) \overset{\Delta}{=} f^0(x,\theta) + \varepsilon J(x)$$

is strictly convex for all θ and $\varepsilon > 0$. Therefore (T,θ) is optimal-solution-continuous in S at θ^0. The second statement of the theorem does not depend on whether Slater's condition is assumed, so the proof from, e.g., [14, Theorem 26.5] applies.

For $j \in P$, denote $f_+^j(x,\theta) = \max\{0, f^j(x,\theta)\}$. Let $\varepsilon > 0$ and for each $j \in P\backslash P^=(\theta)$ let $r_j > 0$. Consider the less constrained perturbed convex program

$$\begin{aligned} &\inf f^0(x,\theta) + \varepsilon J(x) + \sum_{j\in P\backslash P^=(\theta^0)} r_j[f_+^j(x,\theta)]^2 \\ &\text{s.t. } f^j(x,\theta) \leqq 0 \qquad j \in P^=(\theta^0) \end{aligned} \qquad (LC,\theta)$$

Denote its optimal value by $\Phi_{\varepsilon,r}(\theta)$. Programs of the type (LC,θ) were studied in [14] under the assumption that Slater's condition holds (i.e., without the constraints, since $P^=(\theta^0) = \emptyset$) and in the absence of Slater's condition, but for a fixed θ, in [23].

THEOREM 8 If $\tilde{F}(\theta^0)$ is nonempty then (LC,θ) is optimal-solution-continuous in the region S at θ^0 for every $\varepsilon > 0$ and $r_j > 0$, $j \in P\backslash P^=(\theta^0)$. Moreover, if $x(\varepsilon,r)$ denotes the optimal solution of (LC,θ^0) and $\tilde{x}$ is an optimal solution of (2), then

$$\lim_{\varepsilon\to 0^+} \lim_{r\to\infty} x(\varepsilon,r) = \tilde{x} \qquad (3)$$

where $r = \min\{r_j : j \in P\backslash P^=(\theta^0)\}$.

PROOF. Since (P,θ^0) has an optimal solution, the infimum in (LC,θ^0) is achieved. (This follows by [23, Theorem 3].) Clearly, $\tilde{F}(\theta^0)$ is bounded; so (LC,θ) is stable in S at θ^0. The optimal-solution-continuity now follows from the fact that (LC,θ^0) has a unique optimum. Let us prove the second statement. As in the proof of [14, Theorem 26.6], one finds that

$$\delta(||x(\varepsilon,r) - x(\varepsilon)||) \leqq \frac{1}{8\alpha} \sum_{j\in P\backslash P^=(\theta^0)} \frac{u_j^2(\varepsilon)}{r_j} \qquad (4)$$

for some function $\delta(t) > 0$ at $t > 0$ and $\delta(0) = 0$, with $\delta(t) \to 0$ as $t \to 0$; $0 < \alpha < 1$ is a constant. Further, $(x(\varepsilon),u(\varepsilon)) \in F^=(\theta^0) \times R_+^{\text{card } P\backslash P^=(\theta^0)}$ is a saddle point of the "restricted Lagrangian" [23]

$$f^0(x,\theta^0) + \varepsilon J(x) + \sum_{j\varepsilon P\setminus P^=(\theta^0)} u_j f^j(x)$$

on the set $F^=(\theta^0)$. Since Slater's condition holds for the constraints $f^j(x,\theta^0) \leqq 0$, $j \varepsilon P\setminus P^=(\theta^0)$, the vector $u(\varepsilon) = (u_j(\varepsilon))$ is bounded, i.e.,

$$\sup\{||u(\varepsilon)||^2: 0 \leqq \varepsilon \leqq \varepsilon_0\} < \infty \qquad \varepsilon_0 > 0$$

Therefore, the right-hand side in (4) tends to zero as $r \to \infty$. Hence using the properties of $\delta(t)$,

$$\lim_{r\to\infty} x(\varepsilon,r) = x(\varepsilon)$$

This statement, together with Theorem 7, proves (3).

The above results find applications in the study of ordinary convex programs (with differentiable functions)

$$\begin{aligned} &\min f^0(x) \\ &\text{s.t. } f^j(x) \leqq 0 \qquad j \varepsilon P \end{aligned} \tag{P}$$

These programs are considered here as special cases of (P,θ), obtained by specifying $f^0(x,\theta) = f^0(x)$ and $f^j(x,\theta) = f^j(x) - \theta_j$ at $\theta = \theta^0 = 0$. Here we denote $F = \{x \varepsilon R^n: f^j(x) \leqq 0,\ j \varepsilon P\}$, $P^= = \{j \varepsilon P: x \varepsilon F \Rightarrow f^j(x) = 0\}$.

Let x^* denote the optimal solution of (P) having the smallest norm. Then, as a consequence of Theorem 7, the following sample result gives an "error estimate" for the validity of the Kuhn-Tucker condition.

COROLLARY 1 Let x^* be as above. Then for every $\varepsilon > 0$ and $\theta = (\theta_j) \varepsilon R^{\text{card } P^=}$, there exist $x = x(\varepsilon,\theta)$ and $u = u(\varepsilon,\theta) = (u_j(\varepsilon,\theta)) \geqq 0$, $u \varepsilon R^P$ such that

$$\lim_{\varepsilon\to 0^+} \lim_{\theta\to 0^+} x(\varepsilon,\theta) = x^* \tag{5}$$

$$u_j(\varepsilon,\theta) f^j(x(\varepsilon,\theta)) = 0 \qquad j \varepsilon P\setminus P^= \tag{6}$$

$$u_j(\varepsilon,\theta)[f^j(x(\varepsilon,\theta)) - \theta_j] = 0 \qquad j \varepsilon P^= \tag{7}$$

$$\nabla f^0(x(\varepsilon,\theta)) + \sum_{j\varepsilon P} u_j(\varepsilon,\theta) \nabla f^j(x(\varepsilon,\theta)) = -\varepsilon x(\varepsilon,\theta) \tag{8}$$

PROOF. Consider the program (P) perturbed as follows:

$$\inf f^0(x)$$
$$\text{s.t. } f^j(x) \leqq 0 \qquad j \in P \backslash P^=$$
$$f^j(x) - \theta_j \leqq 0 \qquad j \in P^=$$

where $\theta_j > 0$, $j \in P^=$. Since $V_0^= = \{0: 0 \in R^{\text{card } P \backslash P^=}\} \times \{\theta \in R^{\text{card } P^=}: \theta_j > 0, j \in P^=\}$ is here a region of stability, Theorem 7 is applicable with $S = V_0^=$. The corresponding (T,θ) is of the form

$$\inf f^0(x) + \frac{1}{2}\varepsilon||x||^2$$
$$\text{s.t. } f^j(x) \leqq 0 \qquad j \in P \backslash P^=$$
$$f^j(x) - \theta_j \leqq 0 \qquad j \in P^=$$

for some $\varepsilon > 0$. Since Slater's condition holds for the above program, its optimal solution $x = x(\varepsilon,\theta)$ satisfies the Kuhn-Tucker condition, i.e., the Eqs. (6-8). But (T,θ) is optimal-solution-continuous, so

$$\lim_{\theta \to 0^+} x(\varepsilon,\theta) = x(\varepsilon,0) = x(\varepsilon)$$

The relation (5) now holds by the second statement in Theorem 7.

For related asymptotic results the reader is referred to [16] and [24].

VII. CALCULATING THE REGIONS OF STABILITY

In this section we suggest a method for calculating the regions $V_0^=$ and $W_0^=$, or their subsets, emanating from a fixed point θ^0.

METHOD.

i. Find an $x^* \in F(\theta^0)$. Denote by $I = I(\theta) \subset R^m$ those θ's for which $x^* \in F(\theta)$.

ii. Determine a subset $I'(\theta)$ with $\theta^0 \in I'(\theta) \subset I(\theta)$, such that $P(x^*,\theta)$ is independent of θ for $\theta \in I'(\theta)$.

iii. Determine the set $\hat{I} = \hat{I}(\theta) \subset I'(\theta)$ such that $P^=(\theta) \subset P^=(\theta^0)$ for all $\theta \in \hat{I}$. This can be done using the algorithm from [22] for calculating $P^=(\theta)$ as follows:

For every $\theta \in I'$ start with $\Omega(\theta) = \emptyset$. Then increase the set $\Omega(\theta)$ at each iteration in the way described in [22]. Also, at each iteration, calculate the set

$$\hat{I}(\theta) = \{\theta \in I': \Omega(\theta) \subset P^=(\theta^0)\}$$

At the end of the algorithm (after at most card $P^{=}(\theta^0)$ iterations) one obtains the set $\hat{I}$ with the required property.

iv. Determine the two subsets of $\hat{I}$:

$$U_1 = \{\theta \in \hat{I}: P^{=}(\theta) = \emptyset\}$$

$$W_1 = \{\theta \in \hat{I}: P^{=}(\theta) = P^{=}(\theta^0)\}$$

v. *Determination of* $V_0^{=}$: Let U denote those θ's in U_1 with the property that

$$f^j(x,\theta) \leqq 0 \qquad j \in P^{=}(\theta^0)$$

for all $x \in F(\theta^0)$. Then $U \subset V_0^{=}$.

vi. *Determination of* $W_0^{=}$: Calculate $F^{=}(\theta)$ for $\theta \in W_1$. Let $W = \{\theta \in W_1: F^{=}(\theta^0) \subset F^{=}(\theta)\}$. Then $W \subset W_0^{=}$.

EXAMPLE 9. Consider a convex program with the constraints

$$f^1 = (x_1 - 1)^2 + (x_2 - 1)^2 + \theta_1^2 x_3 - \theta_2 \leqq 0$$

$$f^2 = x_1 + x_3^2 + \theta_2 e^{x_4} - 4 \leqq 0$$

$$f^3 = x_2 + \theta_2(e^{-x_3} - 1) - 1 \leqq 0$$

$$f^4 = \theta_3(x_2 - 1) + \theta_4^2(x_1 + x_2 - 2) + \theta_3^2 x_4 \leqq 0$$

$$f^5 = -x_1 - x_2 + 2 \leqq 0$$

perturbed around the point $\theta^0 = (1,1,0,0)^T$. We apply the method:

i. $x^* = (1,1,0,0)^T \in F(\theta)$ for $\theta \in I(\theta) = \{\theta: 0 \leqq \theta_2 \leqq 3\}$.

ii. $P(x^*,\theta) = \{3,4,5\}$, $\theta \in I'(\theta) = \{\theta: 0 < \theta_2 < 3\}$.

iii. We set $\Omega(\theta) = \emptyset$ and consider

$$\sum_{i=3}^{5} y_i \nabla f^i(x^*,\theta) \in [R^4]^* = 0$$

i.e.,

$$\begin{bmatrix} 0 & \theta_4^2 & -1 \\ 1 & \theta_4^2 + \theta_3 & -1 \\ -\theta_2 & 0 & 0 \\ 0 & \theta_3^2 & 0 \end{bmatrix} \begin{bmatrix} y_3 \\ y_4 \\ y_5 \end{bmatrix} = 0$$

This gives:

$$\left.\begin{aligned} &\theta_4^2 y_4 - y_5 &= 0 \\ &y_3 + (\theta_4^2 + \theta_3)y_4 - y_5 &= 0 \\ &-\theta_2 y_3 &= 0 \\ &\theta_3^2 y_4 &= 0 \end{aligned}\right\} \tag{9}$$

First suppose $\theta_3 \neq 0$. Then by (9), for θ near θ^0 so that $\theta_2 \neq 0$, we have $y_3 = y_4 = y_5 = 0$. So for such θ's, the algorithm from [22] terminates and $P^{=}(\theta) = \Omega(\theta) = \emptyset$. Next, suppose $\theta_3 = 0$. From (9),

$$\left.\begin{aligned} &\theta_4^2 y_4 - y_5 &= 0 \\ &y_3 + \theta_4^2 y_4 - y_5 &= 0 \\ &-\theta_2 y_3 &= 0 \end{aligned}\right\} \tag{10}$$

We see that in (10), $y_3 = 0$; if $\theta_4 \neq 0$, y_4 and y_5 can be made positive, while if $\theta_4 = 0$, $y_5 = 0$ and y_4 can be made positive. Suppose $\theta_4 = 0$. Then $\{4\} \subset P^{=}(\theta)$. We set $\Omega = \{4\}$, so $P(x^*,\theta)\backslash\Omega = \{3,5\}$ and

$$D_{f^4}(x^*) = R^4$$

We consider

$$\begin{bmatrix} 0 & -1 \\ 1 & -1 \\ -\theta_2 & 0 \\ 0 & 0 \end{bmatrix} \begin{bmatrix} y_3 \\ y_5 \end{bmatrix} = 0$$

which only has the solution $y_3 = y_5 = 0$, so $P^{=}(\theta) = \{4\}$ when $\theta_3 = \theta_4 = 0$. (In particular, $P^{=}(\theta^0) = \{4\}$.) Now for $\theta_3 = 0$, $\theta_4 \neq 0$, we have $\{4,5\} \subset P^{=}(\theta)$. Thus, from the above discussion, we conclude

$$\hat{I}(\theta) = \{\theta \in I'(\theta)\colon \theta_3 \neq 0 \text{ or } \theta_3 = \theta_4 = 0\}$$

iv. $U_1 = \{\theta\colon 0 < \theta_2 < 3;\ \theta_3 \neq 0\}$

$W_1 = \{\theta\colon 0 < \theta_2 < 3;\ \theta_3 = \theta_4 = 0\}$

v. This step does not provide any suitable θ's; the reason will be apparent from the discussion below.

vi. We are supposing $\theta_2 < 3$, $\theta_3 = \theta_4 = 0$, i.e., those θ's in W_1. We have $P^{=}(\theta) = P^{=}(\theta^0) = \{4\}$, and clearly $F^{=}(\theta) = F^{=}(\theta^0) = R^4$. So

$$W = \{\theta\colon 0 < \theta_2 < 3;\ \theta_3 = \theta_4 = 0\}$$

We now show that in fact this is the only region of stability emanating from θ^0.

Suppose θ is such that $\theta_3 \neq 0$. Now $(1,1,0,1)^T \in F(\theta^0)$; suppose x is near this point. Then

$$f^4(x,\theta) \sim \theta_3^2 > 0 \qquad \text{as } x \notin F(\theta)$$

This means $F(\theta) \not\to F(\theta^0)$ for such θ's, so by Theorem 3 they cannot lie in a region of stability.

Now suppose θ is such that $\theta_3 = 0$ but $\theta_4 \neq 0$. It is easy to check that $(1.1,\ 1.1,\ 1/2,\ 1/2)^T \in F(\theta^0)$; suppose x is near this point. Then

$$f^4(x,\theta) \sim (0.2)\theta_4^2 > 0 \qquad \text{as } x \notin F(\theta)$$

As before, this shows that these θ's cannot lie in a region of stability.

REFERENCES

1. B. Brosowski, On parametric linear optimization, in *Optimization and Operations Research* (R. Henn, B. Korte, and W. Oettli, eds.), 37-44, Lecture Notes in Economics and Mathematical Systems 157, Berlin: Springer-Verlag, 1978.
2. G. B. Dantzig, J. Folkman, and N. Shapiro, On the continuity of the minimum set of a continuous function, *J. Math. Anal. Appl. 17*(1967), 519-548.
3. A. V. Fiacco, Convergence properties of local solutions of sequences of mathematical programming problems in general spaces, *J. Optimization Theory Appl. 13*(1974), 1-12.
4. J. Guddat, Stability in convex quadratic parametric programming, *Math. Operationsforschung Statist.* 7(1976), 223-245.
5. W. Krabs, Stetige Abänderung der Daten bei nichtlinearer Optimierung und ihre Konsequenzen, *Operations Res. Verfahren 1*(1977), 93-113.
6. B. Kummer, Global stability of optimization problems, *Math. Operationsforschung Statist.*, series *Optimization* (1977).
7. R. T. Rockafellar, *Convex Analysis*, Princeton: Princeton University Press, 1970.
8. S. Danø, *Linear Programming in Industry*, New York: Springer-Verlag, 1974.
9. S. M. Robinson, A characterization of stability in linear programming, MRS Tech. Summ. Report 1542, University of Wisconsin, Madison, 1975.

10. D. J. Wilde and C. S. Beightler, *Foundations of Optimization*, Englewood Cliffs, NJ: Prentice-Hall, 1967.

11. N. Williams, *Linear and Non-Linear Programming in Industry*, Pitman, 1967.

12. R. L. Armacost and A. V. Fiacco, Computational experience in sensitivity analysis for nonlinear programming, *Math. Programming 6*(1974), 301-326.

13. D. H. Martin, On the continuity of the maximum in parametric linear programming, *J. Optimization Theory Appl. 17*(1975), 205-210.

14. I. I. Eremin and N. N. Astafiev, *Introduction to the Theory of Linear and Convex Programming* (in Russian), Moscow: Nauka, 1976.

15. A. Ben-Israel, A. Ben-Tal, and S. Zlobec, Optimality conditions in convex programming, in *Survey of Mathematical Programming* (A. Prekopa, ed.), Hungarian Academy of Sciences and Amsterdam: North-Holland, 1979.

16. S. Zlobec and A. Ben-Israel, Perturbed convex programs: Continuity of optimal solutions and optimal values, in *Methods of Operations Research: Proceedings of the III Symposium on Operations Research* (W. Oettli and F. Steffens, eds.), *Verlag Athenäum/Hain/Scriptor/Hanstein 31*(1979), 739-749.

17. A. N. Tikhonov, On incorrectly posed problems of optimal planning, *USSR Comp. Math. Math. Phys. 6*(1966), 114-127.

18. A. N. Tikhonov and V. Y. Arsenin, *Solutions of Ill-Posed Problems*, New York: Wiley, 1977.

19. E. W. Cheney, *Introduction to Approximation Theory*, Hightstown, NJ: McGraw-Hill, 1966.

20. A. Ben-Tal, A. Ben-Israel, and S. Zlobec, Characterization of optimality in convex programming without a constraint qualification, *J. Optimization Theory Appl. 20*(1976), 417-437.

21. H. Wolkowicz, Calculating the cone of directions of constancy, *J. Optimization Theory Appl. 25*(1978), 451-457.

22. R. Abrams and L. Kerzner, A simplified test for optimality, *J. Optimization Theory Appl. 25*(1978), 161-170.

23. S. Zlobec, On the equivalence of convex and unconstrained optimization problems, *Glasnik Matematicki 13*(1978), 397-407.

24. A. V. Fiacco and G. P. McCormick, Asymptotic conditions for constrained minimization, RAC-TP-340, Research Analysis Corporation, McLean, Virginia, November 1968.

Part II APPLICATIONS AND INTERFACES

Chapter 7 PARAMETRIC ANALYSIS IN GEOMETRIC PROGRAMMING: AN INCREMENTAL APPROACH

JOHN J. DINKEL* / Pennsylvania State University, State College, Pennsylvania

GARY A. KOCHENBERGER / Pennsylvania State University, State College, Pennsylvania

DANNY S. WONG / Ohio State University, Columbus, Ohio

ABSTRACT

This paper presents a summary of computational experience with incremental procedures in sensitivity analysis for geometric programs. The incremental procedures are used to approximate discrete changes in parameters by introducing a sequence of small differential changes. The ability of such procedures to generate accurate results is demonstrated by considering changes in the primal coefficients of the objective function and the constraints, the right-hand sides of the constraints, and the exponents of the variables in the objective function.

I. INTRODUCTION

The purpose of this paper is to present a summary of our experiences with sensitivity analysis procedures in geometric programming. In particular, we illustrate the use of an incremental procedure within these sensitivity analyses to provide improved results. This paper focuses on the numerical

---

*Current Affiliation: Texas A&M University, College Station, Texas.

results of procedures whose foundations are reported elsewhere [5,7,9]. While the procedures described here are in the context of geometric programming, they have implications for other approaches to sensitivity analysis as well [1,3,10].

It is important to recognize that most of the sensitivity analysis results in nonlinear programming are based on differential changes in the parameters. However, in most applications, we want to be able to see the effect of discrete changes in the parameters. Thus, we need a mechanism for approximating the results based on differential changes. In addition to generating solutions based on specific parameter changes, we want to be able to compute ranges of parameter values within which the previous approximations are valid. This situation is somewhat analogous to the ranging analysis in linear programming.

The incremental procedure, which approximates a discrete parameter change by a sequence of smaller changes, is used to generate:

i. More accurate numerical solutions using sensitivity procedures;
ii. Approximate bounds or ranges on the parameters for which either currently active constraints become inactive or inactive constraints become active.

It is of interest to note that the incremental approach is an example of a predictor-corrector technique such as Euler's method of integration. The prediction step of the incremental procedure appears to be precisely Davidinko's method [4]*.

II. SENSITIVITY ANALYSIS IN GEOMETRIC PROGRAMMING AND INCREMENTAL PROCEDURES

For a primal geometric program of the form

$$\text{minimize} \quad g_0(t) = \sum_{i=1}^{n_0} c_i \prod_{j=1}^{m} t_j^{a_{ij}}$$

$$\text{subject to} \quad g_k(t) = \sum_{i=m_k}^{n_k} c_i \prod_{j=1}^{m} t_j^{a_{ij}} \leq 1 \tag{1}$$

$$k = 1,\dots,p, \qquad t = (t_1,\dots,t_m) > 0$$

*The authors are grateful to Professor Fiacco for pointing out these similarities and references.

where c_i are positive real numbers, a_{ij} are arbitrary real numbers, and m_k, n_k are indices which consecutively number the primal terms with $n_p = n$, the associated dual geometric program is

$$\text{maximize} \quad v(\delta) = \prod_{i=1}^{n} (c_i/\delta_i)^{\delta_i} \prod_{i=1}^{p} \lambda_k(\delta)^{\lambda_k(\delta)}$$

$$\text{subject to} \quad \sum_{i=1}^{n_0} \delta_i = 1 \tag{2}$$

$$\sum_{i=1}^{n} a_{ij}\delta_i = 0$$

$$j = 1,\ldots,m, \qquad \delta_i \geqq 0, \qquad i = 1,\ldots,n$$

where $\lambda_k(\delta) = \sum_{i=m_k}^{n_k} \delta_i$, $i = 1,\ldots,p$.

The general solution to the constraints of the dual geometric program is

$$\delta = b^{(0)} + \sum_{j=1}^{d} r_j b^{(j)} \geqq 0 \tag{3}$$

where $d = n - m - 1$ is referred to as the degree of difficulty, r_j, $j = 1, \ldots, d$ are new independent variables, and the $b^{(0)}, b^{(1)}, \ldots, b^{(d)}$ are linearly independent vectors. The $b^{(j)}$, $j = 0,1,\ldots,d$ are *constants* and are readily obtainable for large systems [9].

A. Changes in Primal Coefficients

Using this representation and the results of Appendix B in [9], we state Theorem 1 [9], which relates changes in any of the primal coefficients to changes in the optimal solution.

THEOREM 1 *Suppose the primal GP has* $d > 0$ *and rank* $(a_{ij}) = m$. *If the solution to the dual GP has* $\delta^* > 0$ *and if the matrix* $J(\delta)$ *with components*

$$J_{ij}(\delta) = \sum_{q=1}^{n} b_q^{(i)} b_q^{(j)}/\delta_q - \sum_{k=1}^{p} \lambda_k^{(i)}\lambda_k^{(j)}/\lambda_k(\delta) \qquad i,j = 1,\ldots,d \tag{4}$$

is nonsingular at δ^*, *then the functions which give the optimized solution* δ^* *and* $v(\delta^*)$ *in terms of the variable coefficient vector* c *are differentiable on an open neighborhood of* c. *These differentials are:*

$$\frac{dv}{v^*} = \sum_{i=1}^{n} \delta_i^* \frac{dc_i}{c_i} \tag{5}$$

$$d\delta_i = \sum_{j=1}^{d} \left\{ b_i^{(j)} \sum_{\ell=1}^{d} \left[J_{ij}^{-1}(\delta^*) \sum_{i=1}^{n} b_i^{(\ell)} \frac{dc_i}{c_i} \right] \right\} \qquad i = 1,\ldots,n \tag{6}$$

where $J_{ij}^{-1}(\delta^*)$ *represents the components of the inverse of* $J(\delta)$.

This result requires that there be no loose constraints (i.e., all $\delta_i^* > 0$), and hence we assume that the problem has been reformulated so that $\delta^* > 0$. The loose constraints will be included in the analysis following the description of the general methods.

For differential changes, dc_i, that maintain the positivity conditions on all dual variables, the new dual solution is computed as

$$\begin{aligned} \delta_i' &= \delta_i^* + d\delta_i \qquad i = 1,\ldots,n \\ v' &= v^* + v^* \sum_{i=1}^{n} \delta_i^* \frac{dc_i}{c_i} \end{aligned} \tag{7}$$

Once the dual solution has been determined, the new primal solution can be computed using the results of Theorem III.1 in [9].

In order to implement these results, the *differential changes* dc_i/c_i will be approximated by the *difference form* $(c_i' - c_i)/c_i$, where c_i', c_i represent the new and old values, respectively, of the parameters. Such an approximation introduces error into the procedure via the evaluation of J^{-1}. That is, we need to evaluate J^{-1} at the point $(\delta^* + d\delta)$, where $d\delta$ is the change in solution resulting from the change in parameters $(c_i' - c_i)/c_i$. Since $d\delta$ is in general unknown, we avoid the solution of the system (4) under these general conditions by the use of an incremental procedure. This procedure allows for the continued updating of J^{-1} by considering a "small" change about δ^* and computing a new solution

$$\delta_i^1 = \delta_1^* + d\delta_i^1 \tag{8}$$

where

$$d\delta_i^1 = \sum_j b_i^{(j)} \sum_{\ell}^{d} J_{j\ell}^{-1}(\delta^*) \sum_i b_i^{(\ell)} \frac{c_i' - c_i}{c_i} \qquad i = 1,\ldots,n \tag{9}$$

This new solution δ_i^1 is then used as the point about which a new solution is computed for a "small" change; that is, $\delta_i^2 = \delta_i^1 + d\delta_i^2$, and so on.

The incremental procedure has the effect of allowing for updating the evaluation of J^{-1} according to:

$$J_{ij}^{-1}\left(\delta^* + \sum_{t=1}^{T} d\delta^{(t)}\right)$$

where T defines the number of increments.

Our previous work [5,6] has extended these results to the determination of approximate bounds on the allowable changes in a coefficient; that is, the determination of a change $\Delta\delta$ such that $\delta_i' = \delta_i^* + \Delta\delta = 0$. Complete details can be found in [6].

B. Changes in Primal Exponents

While the above results do not allow for changes in the primal exponents, the recent work of Armacost and Fiacco [1], Bigelow and Shapiro [3], and Fiacco [10], and the original work of Fiacco and McCormick [11] do allow for such analysis. Our approach to such analysis within the geometric programming point of view is based on the observation that changes in the primal exponents a_{ij} in (1) correspond to changes in the coefficients in the dual constraints (2). To be consistent with the above references, we write (2) as

$$\begin{aligned} &\text{minimize} && f(x) \\ &\text{subject to} && Ax = b \\ & && x \geqq 0 \end{aligned} \tag{10}$$

where we assume f to be a convex function, and that every solution has positive components, i.e., the nonnegativity constraints are not binding. Under these assumptions the Kuhn-Tucker conditions hold at a solution and, at a positive solution, can be written as [11]:

$$\begin{aligned} \nabla f(x) - A^T\pi &= 0 \\ Ax &= b \end{aligned}$$

where $\nabla f(x) = (\partial f(x)/\partial x_1, \ldots, \partial f(x)/\partial x_n)^T$.

According to the work of Bigelow and Shapiro [3], Fiacco and McCormick [11], and Fiacco [10], under appropriate and well known conditions, the solution to the system of derivatives of the Kuhn-Tucker conditions can be written as

$$\dot{\pi} = \left[A\left(\frac{\partial^2 f}{\partial x^2}\right)^{-1} A^T\right]^{-1} \left[-\dot{A}x + \dot{b} - A\left(\frac{\partial^2 f}{\partial x^2}\right)^{-1}\left(-\frac{\partial^2 f}{\partial p \partial x}\dot{p} + \dot{A}^T\pi\right)\right] \tag{11}$$

$$\dot{x} = \left(\frac{\partial^2 f}{\partial x^2}\right)^{-1} \left[A^T\dot{\pi} - \frac{\partial^2 f}{\partial p \partial x}\dot{p} + \dot{A}^T\pi\right] \tag{12}$$

where π is the vector of Lagrange multipliers for $Ax = b$,

p represents the parameters of the model, i.e., b and A,

$\dot{p}$ represents a change in a parameter, e.g., $\partial b/\partial b_j$, and

$\dot{y}$ represents a change in a variable with regard to the direction $\dot{p}$.

Sufficient conditions [10,11] for the existence of a solution are (i) the second order sufficient condition, (ii) linear independence of the binding constraints, and (iii) strict complementary slackness, along with the twice continuous differentiability of the problem functions. These conditions are required throughout the paper.

Replacing the differential changes by discrete approximations introduces a major source of error via the evaluation of $(\partial^2 f/\partial x^2)$. In order to study changes in the primal exponents, which correspond to changes in A in (10), we set $\dot{p} = \dot{A}$ and $\dot{b} = 0$, for which (10-12) become

$$\Delta\pi = \left[A\left(\frac{\partial^2 f}{\partial x^2}\right)^{-1} A^T\right]^{-1} \left[-\dot{A}x - A\left(-\frac{\partial^2 f}{\partial x^2}\right)^{-1} \dot{A}^T\pi\right] \tag{13}$$

$$\Delta x = \left(\frac{\partial^2 f}{\partial x^2}\right)^{-1} \left[(A^T\dot{\pi} + \dot{A}^T\pi)\right] \tag{14}$$

These results, in conjunction with Theorem III.1 of [9], can then be used to generate the new primal solution for some $\dot{A} \equiv \Delta a_{ij}$.

C. Incremental Procedures

There are several types of incrementing procedures; that is, the way in which $(c_i' - c_i)/c_i$ is chosen. For example, if we have a particular value of c_i' in mind, then the number of increments refers to the number of divisions in the original change. That is, for N increments, the size of each increment is

$$\frac{c_i' - c_i}{c_i}\Big/N = \frac{\Delta}{c_i}\Big/N$$

Thus, *1 increment* means that all computations are made using $J^{-1}(\delta^*)$; *2 increments* means that in (9) $(c_i' - c_i)/c_i$ is replaced by $[(c_i' - c_i)/c_i]/2$, and once δ_i^1 is computed, (9) is recomputed at $J_{jk}^{-1}(\delta^1)$. Clearly, as N increases, J^{-1} is updated more frequently and, as we show in the next section, the results become more accurate.

In another situation, the determination of bounds, a value of c_i' is not known in advance. In this case, we use the term *increment size* to refer to the percentage change in the original coefficient. Thus,

No increments implies that all calculations are made with $J^{-1}(\delta^*)$;
50% increments implies that $\Delta/c_i = 1/2$ for each increment;
10% increments implies that $\Delta/c_i = 1/10$ for each increment;
1% increments implies that $\Delta/c_i = 1/100$ for each increment.

Some other variations are given in [5,6].

Finally, we note that while the above results are given in terms of the c_i, they are equally applicable to the a_{ij} as well.

D. Inactive Constraints and Termination Criteria

The above results presume that all $\delta_i^* > 0$; that is, the problem has been reformulated by discarding any inactive constraints prior to the application of the sensitivity analysis procedures. However, one of the main tenets of such analysis is that constraints that are inactive must remain so. Thus once such constraints are excluded from a problem at the optimal solution, inactive constraints are not encountered again because the problem functions are assumed continuous and the perturbations infinitesimal. See [10,11], for example. Thus, while we do not directly include inactive constraints, we can monitor their values as the parameters change and use this information in the termination of the algorithm.

The indicated conditions lead to the following heuristic stopping rules. Termination of the procedures can occur in three ways as the coefficients vary:

1. A tight constraint becomes loose. This will be detected by some $\delta_i^* = 0$ and denoted by $g_k(t') < 1$.
2. A loose constraint becomes tight. This will be detected by some $g_k(t') \geqq 1$, where $g_k(t^*) < 1$. This is denoted by $g_k(t') > 1$.
3. A maximum limit in the changes is exceeded, denoted by N.S. in Table 1.

Table 1 Ranges of Allowable Changes

Upper Bounds				
	Increment Size			
Coeff.	No	50%	10%	Comment
c_1	24.83	121.42	127.86	N.S.[a]
c_2	57.92	301.41	337.89	N.S.
c_3	86.28	342.94	345.64	N.S.
c_4	275.08	350[b]	370[b]	$g_3(t') > 1$
c_5	13.01	69.23	71.01	N.S.
c_6	177.02	533.23	602.34	N.S.
c_7	79.68	351.37	356.72	N.S.
c_8	32.96	156.15	162.67	N.S.
RHS1	27.50	Not allowed	12[b]	$g_3(t') > 1$
RHS2	300.61	Not allowed	14[b]	$g_3(t') > 1$

Lower Bounds				
	Increment Size			
Coeff.	No	10%	1%	Comment
c_2	0	2[b]	3.8[b]	$g_3(t') > 1$
RHS1	4.63	5[b]	6.2[b]	$g_3(t') > 1$
All others N.S. (LB = 0)				

[a]N.S. denotes "Not Sensitive," meaning the maximum limit has been exceeded.

[b]Last value of coefficient before $g_k(t') > 1$.

For the purposes of this study, we set a limit on allowable increases at 500% and decreases at 99%. The upper limit is arbitrary; the lower limit was set to maintain $c_i' > 0$. If the further behavior of c_i decreasing were of interest, it could be studied by using smaller increments.

III. NUMERICAL RESULTS

We present some computational experience with the preceding methods to demonstrate the ability of the incremental procedures to provide more accurate results. Before detailing these results, two comments are appropriate:

i. Special computational properties of the methods are given in detail elsewhere [6].

ii. There is an obvious tradeoff between number of increments (and hence accuracy) and the cost of the computation.

Within the context of this paper, we are interested in two questions:

1. Can we develop accurate ranges for changes in the parameters? That is, can these procedures be used to predict values for the parameters which will cause currently active constraints to become inactive or vice versa?
2. Can we use the incremental procedures to generate solutions to the primal and dual programs that meet specified levels of accuracy?

The results that follow represent an attempt to draw into one place the illustration of how the incremental procedure can be used to provide more accurate results. While portions of these results have appeared elsewhere, this appears to be the first comprehensive compilation of such results. A problem previously reported in the literature [2] and listed in the appendix is used to illustrate the computational details.

With regard to improved determination of bounds on changes in the various parameters, Table 1 illustrates the impact of the incremental procedure in determining accurate bounds.

In particular, we note that the incremental procedure [6] tends to underestimate the bound. The use of finer increments improves the bound information dramatically. These results are clearly incidated in Table 1. While these provide approximate bounds, we use Table 2 to illustrate the ability of the incremental procedures to identify more accurate bounds.

The results of Table 2 were generated by comparing the optimal solution of various parameter values to the value of the constraints as predicted by the sensitivity analysis procedures. It should be noted that even more accurate bounds could be generated by using still finer increments. However, such precision would increase the computational burden and would probably only be justified in extreme cases.

It should be noted that in Table 1, the changes in the optimal solution were being caused by inactive constraints becoming active. That is, our earlier comment about monitoring inactive constraints is validated. While there were no situations in which an active constraint became inactive, this behavior has been amply illustrated elsewhere [5,6,7,8].

Table 2 Location of Solution Changes for Problem 1

Coefficient	Comments	Value of constraint g_3 according to			
			Sensitivity Analysis		
		Optimal Solution	10% Increments	1% Increments	
c_2 = 6	For c_2 decreasing		.982247	.982175	
4	Table 1 indicates	.998076	.999277	.998696	Actual lower bound between 3.8 and 4
3.8	$g_3(t)$ becomes tight	1[a]		1.000626	
3.6		1[a]		1.00262	
2			1.024107	1.021959	
RHS1 = 11	For RHS1 increasing	.958431	.965272		
12	Table 1 indicates	.997482	1.014785	.996894	Actual upper bound between 12 and 12.1
12.1	g_3 becoming tight	1[a]		1.00278	
12.2		1[a]		1.006894	
13		1[a]	1.067653		
c_4 = 360	For c_4 increasing	.996621	.996617		
367	Table 1 indicates	.998435		.998429	
372	g_3 becoming tight	.999724	1.001562	.999718	
373		.999997		.999976	Actual upper bound between 373 and 374
374		1[a]		1.000233	

[a]Indicates the solution has changed.

The second issue with regard to the use of sensitivity analysis results is the ability of the procedures to generate accurate solution values for the primal and dual variables. This issue is implicit in the previous result in which we monitor the value of the primal constraints. Tables 3 and 4 present some analysis of the accuracy of the primal and dual solutions. These results are obtained by comparing the sensitivity analysis results to the optimal solution at the specified parameter value. The error in the solution is reported as

$$\frac{\text{True - Predicted}}{\text{True}} \times 100\%$$

where "True" corresponds to the variable value as determined by the optimal solution, and "Predicted" corresponds to the variable value as determined by the sensitivity analysis procedures. These tables clearly show the ability of the incremental procedures to control the error, especially in light of large parameter changes.

Comparison of the results of Tables 3 and 4 shows that the primal solution exhibits less error than the dual solution. In addition, the objective function value is relatively insensitive to coefficient changes in that the linear form of (7) provides a good approximation of the true value. While the objective function value displays relative insensitivity, the primal and, in particular, the dual variables indicate the potential error for large coefficient changes.

A. Changes in the Primal Exponents

The use of the sensitivity analysis procedures for studying changes in the exponents of the primal variables is illustrated in the following discussion. The presumption with such changes was that they would have greater impact on the solutions, and thus incremental procedures would be essential. In the analysis that is reported here, the actual changes are kept relatively small (compared to the previous changes), and as will be seen, even small changes require fine increments in order to control the error.

While the procedures used are the same, the mechanism for studying changes in the exponents is a bit different. For a given change in some a_{ij}, say Δa_{ij}, (11) is used to compute $\Delta\pi$, which is then used to compute a new Δt. When used as part of an incremental procedure, the term $\partial^2 f/\partial t^2$ is then evaluated at $t^* + \Delta t$ and the procedure continues. However, in the context of geometric programming, we note that the procedures have the

Table 3 Dual Problem Error Analysis

Parameter Values	c_1+10% (c_1=11)		c_1+100% (c_1=20)			c_1-50% (c_1=5)		c_2+50% (c_2=30)		
Number of Increments	1	10	2	10	100	5	50	2	5	50
δ_1	-.251	-.024	-5.63	-1.002	-.097	-2.963	-.295	-1.325	-.226	-.022
δ_2	-.017	-.0017	-.75	-.147	-.014	-.065	-.004	-2.951	-.539	-.052
δ_3	.016	.0016	.372	.057	.005	.097	.009	.907	.174	.017
δ_4	.052	.0051	1.444	.249	.024	.384	.039	.709	.128	.013
δ_5	-.035	-.0034	-.911	-.15	-.014	-.225	-.022	-.098	-.010	-.0008
δ_6	.056	.0052	1.837	.33	.032	.336	.033	.299	.046	.004
δ_7	.016	.0016	.685	.135	.013	.14	.0138	-1.985	-.366	-.036
δ_8	-.005	-.0003	.352	.089	.009	-.129	-.0145	2.960	.533	.052
δ_9	.322	.031	11.293	1.928	.186	1.715	.172	-.714	-.174	-.018
δ_{10}	.428	.042	17.608	3.025	.293	2.08	.209	.791	.152	.015
δ_{11}	.066	.006	-.251	-.071	-.0093	.688	.074	1.546	.264	.026
δ_{12}	.117	.011	3.213	.539	.052	.738	.074	1.712	.309	.030
δ_{13}	.269	.026	9.561	1.658	.161	1.417	.141	4.066	.706	.068
δ_{14}	.462	.044	16.251	2.713	.262	2.444	.249	.367	.043	.004
δ_{15}	.165	.016	5.022	.866	.084	1.005	.1008	-.428	-.087	-.009
δ_{16}	.262	.026	10.980	1.954	.190	1.243	.121	.142	.0079	.0007
δ_{17}	-.003	-.0007	-1.121	-.276	-.028	.274	.033	2.155	.367	.035
δ_{18}	.099	.009	2.014	.267	.025	.859	.0899	-.006	-.074	-.009
δ_{19}	.0005	.00006	.073	.018	.0018	.006	.0005	-.877	-.159	-.016
δ_{20}	.045	.0049	3.435	.721	.072	.123	.0089	7.462	1.294	.125
v	-.017	-.003	-.722	-.154	-.016	-.059	-.006	-.554	-.119	-.013

Table 4 Primal Problem Error Analysis

Parameter Values	c_1+10%		c_1+100%			c_1-50%		c_2+50%		
Number of Increments	1	10	2	10	100	5	50	2	5	50
t_1	-.046	-.0039	-.472	-.117	-.0091	-.317	0.034	.299	-.257	.006
t_2	.159	.014	1.555	.809	.073	.962	.105	.365	.218	.005
t_3	-.199	-.019	-4.002	-1.186	-.110	-1.098	-.115	.021	-.156	.0003
t_4	.106	.01	2.919	.696	.066	.509	.052	-.258	-.033	-.005
t_5	-.059	-.005	-1.275	-.176	-.016	-.443	-.046	1.044	.494	.020
t_6	.078	.007	2.913	.492	.047	.416	.042	.678	.105	.009
t_7	.012	.009	1.222	.012	.0002	.083	.012	-1.968	.008	-.031
g_0	.01	.0007	.326	.05	.0047	.069	.0076	.0043	-.001	-.0001
g_2	-.043	-.004	-1.239	-.256	-.023	-.248	-.027	.065	.019	.0013
g_6	.013	.0011	-1.417	-.004	-.0003	.143	.017	-.333	-.0645	-.003

c_2+100% (c_2=40)			c_4+200% (c_3=300)		RHS1+10% (b_1=11)			RHS1+20% (b_1=12)	
2	10	100	2	100	1	2	100	5	50
-1.757	-.300	-.029	6.607	1.272	NOT ALLOWED	-.910	-.069	-.973	-.082
-4.004	-.738	-.073	3.384	.614		.372	.027	.393	.029
1.511	.285	.028	-4.482	-.832		-1.422	-.127	-2.118	-.206
1.057	.190	.018	-2.779	-.513		.759	.054	.78	.061
-.062	-.002	-.0005	-1.523	-.271		.461	.054	.843	.095
.326	.049	.005	7.905	1.431		.119	.001	-.615	-.021
-2.865	-.526	-.052	10.74	1.962		-.002	.034	.717	.099
4.776	.869	.085	-.544	-.075		-.370	-.017	-.133	.007
-1.424	-.322	-.033	-2.767	-.546		6.077	.431	9.665	.709
1.303	.242	.024	-4.292	-.809		-.419	-.090	-2.37	-.294
2.112	.357	.035	2.131	.457		7.408	.461	8.225	.499
2.609	.469	.047	-1.334	-.255		1.544	.055	.169	-.003
6.078	1.065	.103	-2.291	-.467		5.172	.349	7.475	.513
.251	.007	-	-2.608	-.056		18.329	1.278	32.317	2.447
-.697	-.147	-.014	-2.093	-.404		1.847	.072	.576	-.077
.003	-.026	-.003	-3.211	-.558		-.638	-.062	-1.104	0.112
3.019	.522	.050	15.03	2.622		2.4	2.16	3.336	.327
-.626	-.210	-.022	17.78	3.097		1.119	.123	2.127	.236
-1.283	-.233	0.023	7.089	1.288		-.213	-.012	-.135	-.004
12.29	2.161	.210	3.484	.683		1.716	.206	3.036	.337
-1.	-.206	-.022	-2.541	-.509		.321	.028	.464	.045

c_2+100%			c_4+200%		RHS1+10% (b_1=11)			RHS1+20% (b_1=12)	
2	10	100	2	100	1	2	100	5	50
.668	-.359	.010	1.898	-.527	NOT ALLOWED	-3.506	0.219	-6.917	-.437
.446	.314	.006	-2.936	2.012		4.644	.299	8.336	.561
.157	-.221	.003	3.777	-1.551		-2.798	-.167	-5.595	-.339
-.432	-.055	-.009	-2.035	.158		3.354	.219	7.232	.490
2.008	.813	.037	-3.364	.004		-1.452	-.093	-2.701	-.181
.979	.138	.013	.258	-.099		2.059	.145	4.277	.308
-3.420	-.026	-.054	-.457	.194		-.756	-.062	-1.147	-.096
-.004	-.003	-	.302	.124		.247	.018	.323	.029
.125	.027	.002	-.348	-.364		-1.363	-.081	-2.862	-.172
-.571	-.088	-.006	-2.878	.922		.346	0.033	-1.018	0.087

same format as the previous procedures. That is, changes in the primal exponents, a_{ij}, are studied via changes in the dual geometric program.

Table 5 presents some analysis of such changes using the incremental procedures. These results indicate the potential for small changes in the exponents to generate much larger changes in the solution values. The encouraging point is that the incremental procedures can control the error in the same way they did with changes in the other parameters. It is interesting to note that for the values representing changes a 10% increase in the exponents of t_1, the one increment procedure erroneously indicated g_2 becoming inactive. However, the incremental procedure with finer increments was able to control this error. An analysis of the dual solution shows that the dual variables associated with g_2 are decreasing and increased parameter changes will eventually force the constraint to become inactive.

Finally, we should note that the results of Table 5 parallel those of Table 4 in that they reflect errors in the primal variables. The actual

Table 5 Error Analysis for Changes in the Primal Exponents

Parameter Values	$a_{11} = 1.1$ (+10%) $a_{21} = -.9$ $a_{41} = -.9$ $a_{51} = 2.2$			$a_{87} = -.9$		$a_{32} = 1.1$ $a_{33} = 1.1$ $a_{34} = 1.1$	
Number of Increments	1	10	100	1	10	1	10
t_1	3.459	.337	.033	-.147	-.017	-.142	-.013
t_2	-1.414	-.129	-.012	-.046	-.004	.283	.025
t_3	1.543	.164	.016	.063	.006	-.057	-.005
t_4	-1.375	-.126	-.012	.177	.018	.275	.025
t_5	- .941	-.080	-.007	-.009	-.001	-.068	-.006
t_6	.969	.117	.012	.236	.024	.212	.019
t_7	- .358	-.036	-.005	.053	.008	-.143	-.013
g_0	- .071	-.005	-.0004	.002	.0005	-.001	0
g_2	1.193	.129	.013	.079	.008	.03	.002
g_6	-1.909	-.207	-.028	-.28	-.029	-.092	-.008

changes are being made in the dual problem, and the primal variables are computed from these changes. An analysis of the dual variables indicates much more accurate results. Thus, it would appear that parameter changes are best studied directly in the problem, primal or dual, in which they occur. The implementation of such a procedure is underway.

IV. SUMMARY

The results of the previous sections indicate the need for and the ability of incremental procedures in parametric analysis. The use of such procedures provides a mechanism for generating more accurate results in terms of controlling the potential error. These procedures should enhance the applicability of sensitivity analysis procedures in nonlinear programming by allowing for the use of larger discrete changes in the parameters, relative to a given allowable error, or for reducing the error introduced by approximations using a specified range of parameter changes.

APPENDIX

PROBLEM 1. Source: Beck and Ecker [2], problem 13.

			a_{ij}						
	i	c_i	$j = 1$	2	3	4	5	6	7
g_0	1	10	1	-1	-1	1	0	0	0
	2	20	-1	0	0	-1	1	1	0
	3	30	0	1	1	1	0	0	0
	4	100	-1	-1	-1	-1	-1	-1	-1
	5	5	2	2	1	0	1	3/2	2
	6	50	0	0	-1/2	-1/2	-1/2	0	0
	7	25	0	0	2	2	-1	-1	-1
	8	10	0	0	1/2	1/2	1	1	1
g_1	9	.1	2	2	1	0	0	0	0
	10	.05	0	0	0	1	1/2	0	0
	11	.15	0	0	0	0	0	1/2	1/2
g_2	12	.1	1	0	0	1	0	0	1
	13	.05	1	-1	-1	0	1	1	1/2
	14	.05	0	2	2	-1	0	0	0
	15	.15	-1/2	-.3	1	0	1/2	0	0
	16	.1	0	0	0	0	1	1	0
	17	.1	0	0	0	2	0	0	0
	18	.2	1	1	1	0	0	0	0

$$g_3(t) = \frac{1}{10} \sum_{j=1}^{7} t_j \leqq 1 \qquad g_4(t) = \frac{1}{50} \sum_{j=1}^{7} t_1 t_j \leqq 1$$

$$g_5(t) = \frac{1}{100} \sum_{j=1}^{6} t_j t_{j+1}^{-1} \leqq 1 \qquad g_6(t) = \frac{1}{10} \sum_{j=1}^{5} t_j t_{j+2}^{-2} \leqq 1$$

$$g_7(t) = \frac{1}{50} \sum_{j=1}^{5} t_{j+2} t_j^{-1/2} \leqq 1 \qquad t_j > 0 \qquad j = 1,\ldots,7$$

Optimal Solution:

$t_1 = 1.341865$	$t_5 = 3.148992$	$g_0 = 178.47791$	$g_4 = .24789$
$t_2 = .993245$	$t_6 = .403874$	$g_1 = .35547$	$g_5 = .11768$
$t_3 = .870469$	$t_7 = 1.547682$	$g_2 = 1$	$g_6 = 1$
$t_4 = .923592$		$g_3 = .92289$	$g_7 = .12685$

REFERENCES

1. R. L. Armacost and A. V. Fiacco, Computational experience in sensitivity analysis for nonlinear programming, *Math. Programming 6*(1974), 301-326.

2. P. A. Beck and J. G. Ecker, Some computational experience with a modified convex simplex algorithm for GP, *J. Optimization Theory Appl. 14*(1974).

3. J. H. Bigelow and N. Z. Shapiro, Implicit function theorems with mathematical programming and for systems of inequalities, *Math. Programming 6*(1974), 141-156.

4. D. Davidinko, On a new method of numerically integrating a system of nonlinear equations, *Dokl. Akad. Nauk. SSSR 88*(1953), 600-604.

5. J. J. Dinkel and G. A. Kochenberger, On sensitivity analysis in geometric programming, *Operations Res. 20*(1977), 155-163.

6. J. J. Dinkel, G. A. Kochenberger, and S. N. Wong, Sensitivity analysis procedures for geometric programs: Computatonal aspects, *ACM Trans. Math. Software 4*(1978), 1-14.

7. J. J. Dinkel and G. A. Kochenberger, An implementation of some implicit function theorems with applications to sensitivity analysis, *Math. Programming 15*(1978), 261-267.

8. J. J. Dinkel and G. A. Kochenberger, Sensitivity analysis and the entropy maximization problem, paper presented at ORSA/TIMS Meeting, Atlanta, November 7-9, 1977.

9. R. J. Duffin, E. L. Peterson, and C. Zener, *Geometric Programming*, New York: Wiley, 1967.

10. A. V. Fiacco, Sensitivity analysis for nonlinear programming using penalty methods, *Math. Programming 10*(1976), 287-311.

11. A. V. Fiacco and G. P. McCormick, *Nonlinear Programming: Sequential Unconstrained Minimization Techniques*, New York: Wiley, 1968.

Chapter 8 PRELIMINARY SENSITIVITY ANALYSIS OF A STREAM POLLUTION ABATEMENT SYSTEM*

ANTHONY V. FIACCO and ABOLFAZL GHAEMI† / The George Washington University, Washington, D.C.

I. INTRODUCTION

The present paper reports on the sensitivity analysis study of a geometric programming model‡ of a water pollution control system applied by J. G. Ecker [7] to define and solve three geometric programming problems involving data for the upper Hudson River. We have formulated and solved the convex equivalents of the above problems using the SUMT code [12], and have developed a computer routine to calculate the coefficients involved in the dissolved oxygen constraints in terms of the model parameters. These results are reported in [9]. By sensitivity analysis is meant an analysis of the effect on the optimal objective function value and on an optimal solution point of small perturbations in the model parameters. The importance of such an analysis in real world optimization problems

---

*Research supported in part by National Science Foundation Grant ENG-7906104 and in part by Office of Naval Research Contract N00014-75-C-0729.

†Current affiliation: Reservoir Engineering Department, Iranian Marine International Oil Company, Teheran, Iran.

‡The general model was apparently originally proposed by Charnes and Gemmell. See Section V of this paper.

cannot be overstated. It provides the model maker and user with invaluable information regarding the functional relationship between a solution and the design parameters. This has many potential applications. For example, identification of those parameters having the most significant impact on the optimal solution can provide a basis for developing educated guidelines for taking appropriate and efficient action toward effecting parameter changes that will give an optimal marginal improvement of system performance.

Similar studies have been conducted by Armacost and Fiacco on a variety of problems, including a cattle feed problem [2] and a multi-item inventory problem [3]; and by the authors on a nonlinear structural design problem [10].

The theoretical basis for the approach taken here to calculate solution sensitivity was originally given in the work of Fiacco and McCormick [11]. Fiacco subsequently generalized this theory and established a theoretical basis for utilizing a penalty function method to estimate the sensitivity information of a local solution and its associated Lagrange multipliers, for a large class of nonlinear programming problems, with respect to general parametric variations of the problem functions [8]. A computational algorithm, "SENSUMT," was devised [4] to implement this method, and subsequently integrated with SUMT [12], evolving through a sequence of refinements. The latest revision by Armacost [1] is filed in The George Washington University Center for Academic and Administrative Computing. SENSUMT provided the main tool for the present study.

The sensitivity analysis is conducted with respect to all the model parameters for the fixed dissolved oxygen level policy, which yielded the minimum annual pollution control cost for an upper Hudson River data base, among three environmental control policies applied [7,9]. It is concluded that the optimal waste treatment cost is most sensitive to the maximum allowable oxygen deficit. It is shown that a relaxation of one percent of this requirement, which amounts to about .0218 mg/ℓ, would save approximately $10,270 per year in waste treatment costs. It is also shown that the optimal cost is quite sensitive to several parameters involved in the dissolved oxygen deficit equation [9, p. 12, Equation (1)], and in particular to the fraction of the river bottom covered with sludge, the oxygen uptake rate per unit of river bottom surface area, and the hydraulic radius of the river. The solution is also rather sensitive to certain Biochemical Oxygen Demand (BOD) removal requirements (by primary and secondary treatment), and moderately sensitive to the deoxygenation constant, BOD concentration of the effluent before treatment, the volume of the effluent released into the river, and the river's flow rate. The optimal treatment cost appears relatively insensitive to the remaining model parameters, at least for small data changes.

II. PROBLEM STATEMENTS AND SOLUTIONS

We have chosen to conduct the sensitivity analysis for the policy mandating a fixed dissolved oxygen requirement, using an upper Hudson River data base. Although the details concerning the formulation and solution of this problem (Problem P_1) were presented in our previous work [9], for completeness we give the original GP formulation, and the respective optimal solutions that were obtained. The general problem is to determine treatment levels of various system components that minimize the total annual treatment cost, subject to constraints on maximum allowable oxygen deficits and component treatment levels that may vary between reaches.

A depiction of the system treatment facility configuration is given in Figure 1, and the associated problem formulations follow.

PROBLEM P_1. Corresponding to a "fixed dissolved oxygen requirement policy" of 6.2 mg/ℓ along the river. (Dimensions: 22 variables, 42 constraints.)

$$\begin{aligned}
\text{Minimize } F = {} & 19.4\, t_{11}^{-1.47} + 86.0\, t_{12}^{-.38} + 152\, t_{13}^{-.27} \\
& + 19.4\, t_{21}^{-1.47} + 16.8\, t_{22}^{-1.66} + 27.4\, t_{23}^{-.63} + 179\, t_{24}^{-.37} \\
& + 19.4\, t_{31}^{-1.47} + 16.8\, t_{32}^{-1.66} + 91.5\, t_{33}^{-.30} + 120\, t_{34}^{-.33} \\
& + 19.4\, t_{41}^{-1.47} + 45.9\, t_{42}^{-.45} + 179\, t_{43}^{-.37} \\
& + 19.4\, t_{51}^{-1.47} + 16.8\, t_{52}^{-1.66} + 91.5\, t_{53}^{-.30} + 152\, t_{54}^{-.27} \\
& + 19.4\, t_{61}^{-1.47} + 16.8\, t_{62}^{-1.66} + 27.4\, t_{63}^{-.63} + 179\, t_{64}^{-.37}
\end{aligned}$$

subject to:

a. Dissolved oxygen deficit constraints, i.e., dissolved oxygen deficit for all reaches $\leqq$ 2.18 mg/ℓ.

(1) $u_{11}t_{11}t_{12}t_{13} \leqq 1$

(2) $u_{21}t_{11}t_{12}t_{13} + u_{22}t_{21}t_{22}t_{23}t_{24} \leqq 1$

(3) $u_{31}t_{11}t_{12}t_{13} + u_{32}t_{21}t_{22}t_{23}t_{24} + u_{33}t_{31}t_{32}t_{33}t_{34} \leqq 1$

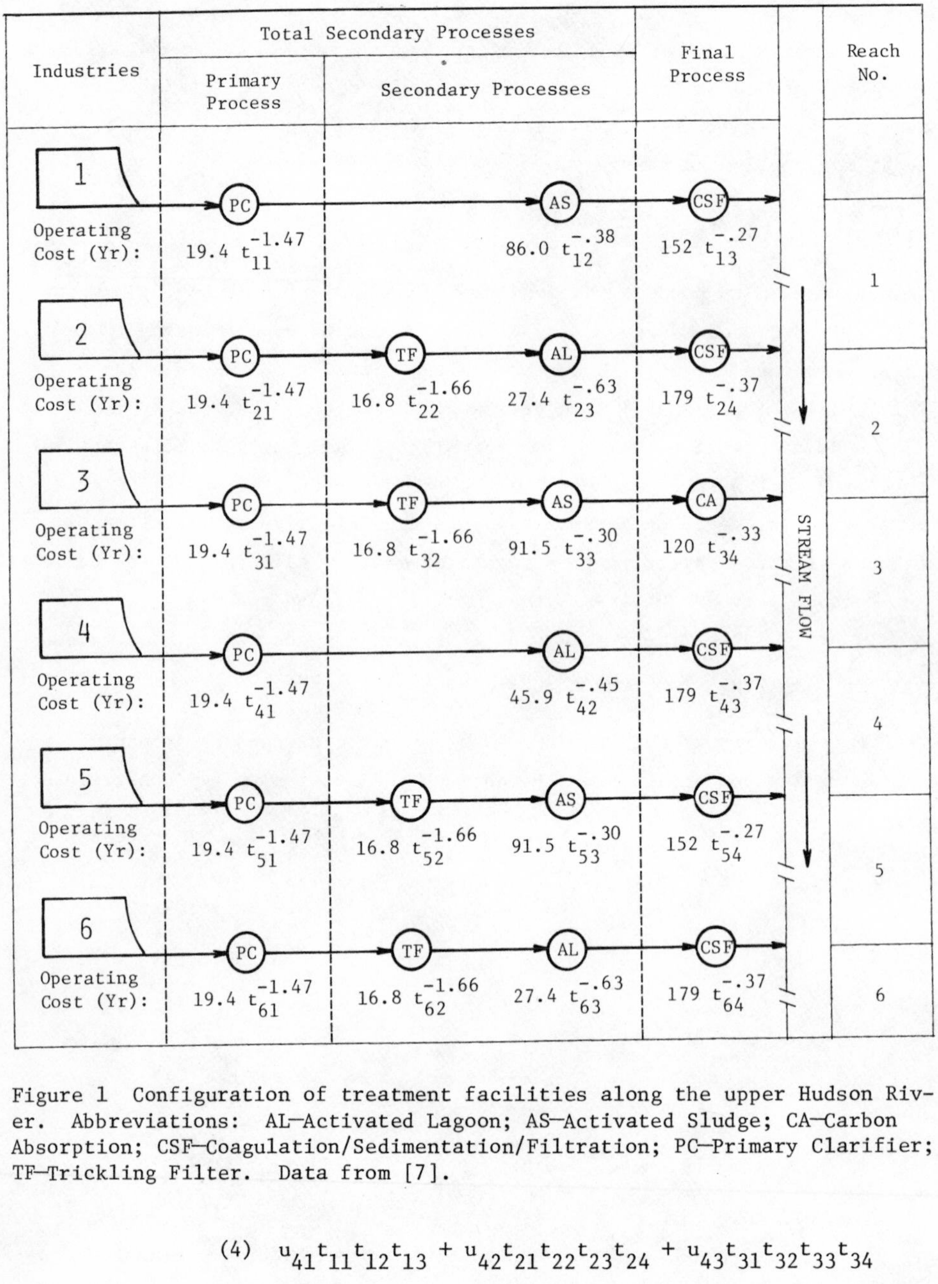

Figure 1 Configuration of treatment facilities along the upper Hudson River. Abbreviations: AL—Activated Lagoon; AS—Activated Sludge; CA—Carbon Absorption; CSF—Coagulation/Sedimentation/Filtration; PC—Primary Clarifier; TF—Trickling Filter. Data from [7].

$$(4)\quad u_{41}t_{11}t_{12}t_{13} + u_{42}t_{21}t_{22}t_{23}t_{24} + u_{43}t_{31}t_{32}t_{33}t_{34} + u_{44}t_{41}t_{42}t_{43} \leqq 1$$

$$(5)\quad u_{51}t_{11}t_{12}t_{13} + u_{52}t_{21}t_{22}t_{23}t_{24} + u_{53}t_{31}t_{32}t_{33}t_{34} + u_{54}t_{41}t_{42}t_{43} + u_{55}t_{51}t_{52}t_{53}t_{54} \leqq 1$$

(6) $u_{61}t_{11}t_{12}t_{13} + u_{62}t_{21}t_{22}t_{23}t_{24} + u_{63}t_{31}t_{32}t_{33}t_{34}$
$+ u_{64}t_{41}t_{42}t_{43} + u_{65}t_{51}t_{52}t_{53}t_{54}$
$+ u_{66}t_{61}t_{62}t_{63}t_{64} \leqq 1$

b. Constraints on combination of processes.

(7) $.10\, t_{11}^{-1}\, t_{12}^{-1} \leqq 1$ at most 90% removal by total secondary treatment on reach 1

(8) $.15\, t_{21}^{-1}\, t_{22}^{-1}\, t_{23}^{-1} \leqq 1$ at most 85% removal by total secondary treatment on reach 2

(9) $.15\, t_{31}^{-1}\, t_{32}^{-1}\, t_{33}^{-1} \leqq 1$ at most 85% removal by total secondary treatment on reach 3

(10) $.15\, t_{41}^{-1}\, t_{42}^{-1} \leqq 1$ at most 85% removal by total secondary treatment on reach 4

(11) $.15\, t_{51}^{-1}\, t_{52}^{-1}\, t_{53}^{-1} \leqq 1$ at most 85% removal by total secondary treatment on reach 5

(12) $.15\, t_{61}^{-1}\, t_{62}^{-1}\, t_{63}^{-1} \leqq 1$ at most 85% removal by total secondary treatment on reach 6

(13) $1.4286\, t_{61}t_{62} \leqq 1$ at least 30% removal by first two components in reach 6

c. Operating range constraint for components of treatment facilities.

(14) $1.25\, t_{41} \leqq 1$ at least 20% removal by first component in reach 4

d. Redundant constraints (added to prevent numerical overflow).

(15) $e^{-80}\, t_{11}^{-1}\, t_{12}^{-1}\, t_{13}^{-1} \leqq 1$

(16) $e^{-80}\, t_{21}^{-1}\, t_{22}^{-1}\, t_{23}^{-1}\, t_{24}^{-1} \leqq 1$

(17) $e^{-80}\, t_{31}^{-1}\, t_{32}^{-1}\, t_{33}^{-1}\, t_{34}^{-1} \leqq 1$

(18) $e^{-80}\, t_{41}^{-1}\, t_{42}^{-1}\, t_{43}^{-1} \leqq 1$

(19) $e^{-80}\, t_{51}^{-1}\, t_{52}^{-1}\, t_{53}^{-1}\, t_{54}^{-1} \leqq 1$

(20) $e^{-80}\, t_{61}^{-1}\, t_{62}^{-1}\, t_{63}^{-1}\, t_{64}^{-1} \leqq 1$

e. Natural constraints.

(21) — (42) $0 < t_{ij} \leqq 1$ $i = 1,6;\ j = 1,m_i$

The coefficients u_{ij} in the constraints 1-6 are complicated functions [9, Appendix 1] of the maximum allowable dissolved oxygen deficit and the physical parameters characterizing the reaches. These coefficients may be evaluated for any problem parameter values via a computer program developed by the authors [9, Appendix 2]. The values of the coefficients obtained using the upper Hudson River data are listed in Appendix 3 with the respective data base.

OPTIMAL SOLUTION OF PROBLEM P_1

Optimal Treatment Levels[a]

t^*_{ij}	j = 1	j = 2	j = 3	j = 4
i = 1	.7236	.4095	.6617	--
i = 2	.7729	.7854	.2471	.4343
i = 3	.8640	.8668	.4303	1.0000
i = 4	.8000	.3229	1.0000	--
i = 5	1.0000	1.0000	1.0000	1.0000
i = 6	.7729	.7854	.2471	.9551

Note: Annual waste treatment cost = $1,837,851; binding constraints: (6, 8, 12, and 14).

[a]Using a fixed dissolved oxygen standard of 6.2 mg/ℓ along the entire stream.

Optimal Plant BOD Removal Levels and Costs[a]

Reach No.	Total BOD Removal (%)	Cost/Year ($)	Redundant Components
1	80.39	321,870	None
2	93.48	363,230	None
3	67.77	163,190	CA[b]
4	74.17	103,260	CSF[b]
5	0.00	0	All
6	85.67	301,590	None
		1,253,140 (Total)	

[a]Using a fixed dissolved oxygen standard of 6.2 mg/ℓ along the entire stream.

[b]See Figure 1 for definitions.

III. SENSITIVITY ANALYSIS

A list and description of all model parameters are given in Appendix 1. The analysis was conducted with respect to all the model parameters except for parameter $\overline{PP}$, which is not involved in the formulation of Problem P_1. The objective is to assess the relative impact of perturbations of the parameters on the optimal solution vector and cost. Having done this, a number of inferences are made concerning guidelines for improving the treatment operation.

Sensitivity information for the optimal value function $f^*(\varepsilon)$ is calculated in Table 1. We have tabulated the linear estimates $\partial f^*/\partial \varepsilon_n$ of changes in $f^*(\varepsilon)$ associated with unit changes in a given parameter ε_n, and changes due to a one percentage increase in ε_n, i.e., $.01\varepsilon_n \partial f^*/\partial \varepsilon_n$. For completeness, the sensitivities $\partial t^*_{ij}/\partial \varepsilon_n$ for the solution vector are tabulated in Appendix 2.

Furthermore, assuming that the parameters involved in the oxygen deficit constraints (constraints numbered 1-6 in Problem P_1) can be varied independently from reach to reach, we can sum the scaled sensitivities in the different reaches to find a linear estimate of the change in $f^*(\varepsilon)$ across all reaches resulting from the simultaneous increase in each reach by one percent of each of the involved parameters. This allows us to calculate the *overall* effect, along the entire length of stream involved, of each model parameter on the optimal annual treatment cost. The results are given in the second column of Table 2. The last column gives the ranks of the absolute values of the respective entries in column 2, ordering these values from the largest to the smallest.

A multitude of inferences can be drawn from the solution vector sensitivity information given in Appendix 2 and the optimal solution value sensitivity given in Tables 1 and 2. The solution vector information can be used to predict optimal system component treatment levels that will be required for any combination of small changes in the parameters, also pinpointing the possible need for adding new components (e.g., if predicted requirements on a given component are deemed excessive) and for removing others (e.g., if predicted requirements are below a given level).

Table 1 Optimal Value Function Sensitivity Results[a]

Parameter Number (n)	Parameter (ε_n)	Parameter Value	Change Due to One Unit Increase in Parameter Value ($1000's)	Change Due to 1% Increase in Parameter Value ($1000's)
1	V_{R_0}	1162.000	-0.22	-2.556
2	L_{b_0}	1.000	114.82	1.148
3	D_{b_0}	1.000	101.82	1.018
4	K_1	0.126	278.52	0.351
5	K_2	0.126	60.92	0.077
6	K_3	0.126	116.74	0.147
7	K_4	0.126	746.07	0.940
8	K_5	0.126	168.49	0.212
9	K_6	0.126	994.08	1.252
10	r_1	0.080	-157.43	-0.126
11	r_2	0.330	27.94	0.092[b]
12	r_3	0.224	-39.51	-0.088
13	r_4	0.216	-349.96	-0.756
14	r_5	0.216	-80.53	-0.152
15	r_6	0.250	-202.36	-0.506
16	t_1	4.000	5.62	0.225
17	t_2	0.520	32.49	0.169
18	t_3	0.700	8.37	0.058
19	t_4	2.620	7.02	0.184
20	t_5	0.370	16.20	0.060
21	t_6	0.950	78.57	0.746
22	F_1	0.700	51.62	0.361
23	F_2	0.700	70.57	0.494
24	F_3	0.700	54.37	0.381
25	F_4	0.700	287.64	2.013
26	F_5	0.700	58.49	0.409
27	F_6	0.700	225.26	1.577
28	Q_1	3.280	11.01	0.361
29	Q_2	3.280	15.06	0.494

Table 1 (Continued)

Parameter Number (n)	Parameter (ε_n)	Parameter Value	Change Due to One Unit Increase in Parameter Value ($1000's)	Change Due to 1% Increase in Parameter Value ($1000's)
30	Q_3	3.280	11.60	0.381
31	Q_4	3.280	61.38	2.013
32	Q_5	3.280	12.48	0.409
33	Q_6	3.280	48.07	1.577
34	K_{s_1}	2.000	18.07	0.361
35	K_{s_2}	2.000	24.70	0.494
36	K_{s_3}	2.000	19.03	0.381
37	K_{s_4}	2.000	100.67	2.013
38	K_{s_5}	2.000	20.47	0.409
39	K_{s_6}	2.000	78.84	1.577
40	R_{w_1}	4.880	-7.40	-0.361
41	R_{w_2}	2.440	-20.25	-0.494
42	R_{w_3}	3.410	-11.16	-0.380
43	R_{w_4}	3.380	-59.57	-2.013
44	R_{w_5}	3.200	-12.79	-0.409
45	R_{w_6}	2.900	-54.37	-1.576
46	V_{E_1}	16.000	2.64	0.422
47	V_{E_2}	32.500	2.57	0.835
48	V_{E_3}	6.400	5.32	0.340
49	V_{E_4}	41.500	0.71	0.295
50	V_{E_5}	48.500	0.12	0.058
51	V_{E_6}	20.000	3.28	0.656
52	E_1	148.000	0.31	0.459
53	E_2	436.000	0.21	0.915
54	E_3	179.000	0.20	0.358
55	E_4	38.000	0.97	0.369
56	E_5	4.830	2.25	0.109
57	E_6	661.000	0.10	0.661
58	S_1	2.180	0.00	0.000

Table 1 (Continued)

Parameter Number (n)	Parameter (ε_n)	Parameter Value	Change Due to One Unit Increase in Parameter Value ($1000's)	Change Due to 1% Increase in Parameter Value ($1000's)
59	S_2	2.180	0.00	0.000
60	S_3	2.180	0.00	0.000
61	S_4	2.180	0.00	0.000
62	S_5	2.180	0.00	0.000
63	S_6	2.180	-471.07	-10.270
64	$\underline{P}_6$	0.300	0.00	0.000
65	$\bar{P}_1$	0.900	0.00	0.000
66	$\bar{P}_{2-6}$	0.850	-494.98	-4.207
67	$\underline{PP}_4$	0.200	3.18	0.006

[a]See Appendix 1 for parameter definitions.

[b]This quantity appears inconsistent, but we could not uncover any error in its calculation.

IV. OBSERVATIONS AND CONCLUSIONS

Based on Tables 1 and 2, the following conclusions are valid.

1. The order of the parameters, arranged according to the decreasing order of the respective absolute magnitude of the changes in the optimal annual treatment cost resulting from a one percent parameter change, is as follows: S_6, (F, Q, K_s, R_w), $\bar{P}_{2-6}$, K, E, V_E, V_{R_0}, r, t, L_{b_0}, D_{b_0}, $\underline{PP}_4$, (S_1 - S_5, $\underline{P}_6$, $\bar{P}_1$), where the parentheses indicate ties.
2. The optimal annual treatment cost decreases with increases in S_6, R_w, $\bar{P}_{2-6}$, V_{R_0}, and r, is insensitive to changes in S_1 - S_5, $\underline{P}_6$ and $\bar{P}_1$, and increases with increases in the remaining parameters.
3. The magnitude of the significant decreases in the optimal annual treatment cost ranges from $10,270 per year for a one percent increase in the parameter S_6 to $1,081 per year for a one percent decrease in the parameter D_{b_0}.
4. Although the maximum allowable oxygen deficit S_6 clearly dominates the other parameters taken individually, it is interesting to note that the cumulative effects of simultaneously decreasing F, Q, and K_s by

Table 2 Changes Summed Across Reaches in the Minimum Annual Waste Treatment Cost Due to a 1% Increase in the Model Parameters

Parameter	Change in Optimal Value Function Corresponding to a 1% Increase in the Given Parameter Type Across All Reaches	Rank
V_{R_0}	-2.556	7
L_{b_0}	1.148	10
D_{b_0}	1.081	11
K	2.979	4
r	-1.536	8
t	1.442	9
F	5.235	2
Q	5.235	2
K_s	5.235	2
R_w	-5.235	2
V_E	2.606	6
E	2.871	5
$S_1 - S_5$	0.000	13
S_6	-10.270	1
$\underline{P}_6$	0.000	13
$\bar{P}_1$	0.000	13
$\bar{P}_{2-6}$	-4.207	3
$\underline{PP}_4$	0.006	12

one percent of their given values, all such changes associated with removing sludge from the river bottom, would result in an estimated annual treatment cost reduction of $15,705, a decrease of about 50% below the reduced cost noted for a one percent increase in S_6. Thus, depending on cost trade-offs, of course, sludge removal would appear to be a good possibility for realizing an optimal marginal improvement.

The preceding point suggests that some parameters may be altered simultaneously by a given change, a fact that should obviously be considered in a cost analysis involved with design modifications. Obviously any attempt to alter or measure the parameters more accurately would generally involve some cost. Such changes or efforts are economically attractive only if it is determined that the cost incurred will be less than the expected gain. Such an assessment would require results such as those provided by the present study. Sensitivity information, coupled with judicious interpretation by users conversant with the application of the model, could provide invaluable insights and guidelines for determining the most cost-effective changes in environmental control policies and systems design parameters, providing a more complete basis for determining the economic implications of making these changes. Indeed, such information would appear to be crucial not only for assessing the impact of changes *without re-solving* the problem with new data and new parameter values, but even for interpreting the significance of a "solution."

V. POSTSCRIPT: COMMENT ON THE DEVELOPMENT OF THE MODEL

While this paper was in its final stages of preparation, we learned [6], in response to our previous paper [9], that A. Charnes and R. E. Gemmell originally formulated a model of similar character [5]. Their model was subsequently included in the 1973 Istanbul NATO conference proceedings. The definition of the model variables, form of the cost function, and structure of the dissolved oxygen deficit constraints utilized by Ecker conform to those introduced in these referenced works. The earlier models apparently assume that the coefficients involved in the oxygen deficit equation relationships are values specified at the outset. One of the important innovations introduced by Ecker is the functional dependence of these coefficients (u_{ij} in Problem P_1) on S_i, the maximum allowable oxygen deficit in reach i, and on the numerous parameters (given in Appendix 1) that are involved in defining the dissolved oxygen deficit equation that is crucial in characterizing the state of the stream. These relationships were developed in some detail in the Ecker paper [7] and in our earlier paper [9].

It is also relevant to note that a more detailed analysis of the model and the optimal value sensitivity information, based both on [9] and on the results of this paper, will be developed in another paper intended for journal publication.

APPENDIX 1 Listing and Description of the Parameters Involved in the Formulation of the Model

Parameters representing physical measurements involved in the dissolved oxygen deficit equation of Problem P_1:

V_{R_0} flow rate of the river before entering reach 1, in 10^6 gallons/day
L_{b_0} initial BOD level of the stream before entering reach 1, in mg/ℓ
D_{b_0} initial oxygen deficit of the stream before entering reach 1, in mg/ℓ
K deoxygenation constant, in day^{-1}
r reaeration constant, in day^{-1}
t flow time along the reach, in days
F fraction of river bottom covered with sludge
Q a coefficient in the sludge term, determined empirically in day^{-1}
K_s oxygen uptake rate per unit area of stream bottom surface, in gm/m^2/day
R_w hydraulic radius of the river cross section, in meters
V_E volume of the effluent released into the river, in 10^6 gallons/day
E BOD concentration of effluent, before treatment, in mg/ℓ.

Aside from the above parameters that are involved in the oxygen sag equation of each reach, the following parameters will also enter into the formulation of the model at various stages.

Parameters representing quantities determined by management decisions or feasibility considerations:

S maximum allowable oxygen deficit, in mg/ℓ
$\underline{P}$ minimum required or feasible BOD removal by a specified sequence of treatment components in a given reach; a fraction
$\overline{P}$ maximum required or feasible BOD removal by a specified sequence of treatment components in a given reach; a fraction
$\underline{\underline{PP}}$ minimum required or feasible BOD removal by a specified single treatment component in a given reach; a fraction
$\overline{\overline{PP}}$ maximum required or feasible BOD removal by a specified single treatment component in a given reach; a fraction.

In Table 1, parameters $\overline{P}_1$, $\overline{P}_{2-6}$, and $\underline{P}_6$ and $\underline{\underline{PP}}_4$ have the following meanings:

$\underline{P}_6$ minimum required BOD removal by the first two treatment components in reach 6
$\overline{P}_1$ maximum required BOD removal by the total secondary treatment

components in reach 1

$\bar{P}_{2-6}$ maximum required BOD removal by the total secondary treatment components in reaches 2, 3, 4, 5, and 6. (We have taken $\bar{P}_{2-6} = \bar{P}_2 = \bar{P}_3 = \ldots = \bar{P}_6$ in this analysis, although our general model allows for independent variation of $\bar{P}_2 - \bar{P}_6$.)

$\underline{PP}_4$ minimum required BOD removal by the first treatment component in reach 4.

APPENDIX 2* Sensitivities of the Problem Variables t^*_{ij}, the Fraction of BOD Remaining After Treatment Component j in Reach i at the Optimal Solution Point of Problem P_1, with Respect to the Problem Parameters ε_n

Table A.1

Variable	Parameter	
	$\varepsilon_2 = L_{b_0}$ = 1 mg/ℓ	$\varepsilon_3 = D_{b_0}$ = 1 mg/ℓ
t_{13}	-.1649	-.1462
t_{24}	-.1744	-.1546
t_{33}	-.1387	-.1230
t_{42}	-.1231	-.1092
t_{64}	-.3833	-.3399

*Sensitivities with absolute value less than 0.1, i.e., $|\partial t_{ij}/\partial \varepsilon_n| < 0.1$, though calculated, are not tabulated here. This accounts for blank entries and derivatives not appearing in the tables that follow.

Table A.2

Variable	Parameter					
	$\varepsilon_4 = K_1 =$ .126 day^{-1}	$\varepsilon_5 = K_2 =$ .126 day^{-1}	$\varepsilon_6 = K_3 =$ .126 day^{-1}	$\varepsilon_7 = K_4 =$ .126 day^{-1}	$\varepsilon_8 = K_5 =$ .126 day^{-1}	$\varepsilon_9 = K_6 =$ .126 day^{-1}
t_{11}				-.1901		-.1569
t_{12}	-.1762			-.4161		-.3434
t_{13}	-.4006		-.1674	-.9430	-.2158	-.7809
t_{24}	-.4233	-.1225	-.2285	-1.3720	-.3054	-1.0615
t_{31}	-.1362			-.4666	-.1035	-.3573
t_{32}	-.1211			-.4146		-.3174
t_{33}	-.3326		-.1879	-1.1389	-.2525	-.8720
t_{42}	-.3007			-1.1312	-.2493	-.8518
t_{64}	-.9306	-.1557	-.2409	-1.0991	-.2749	-.7223

Table A.3

Variable	Parameter					
	$\varepsilon_{10} = r_1 = .080\ day^{-1}$	$\varepsilon_{11} = r_2 = .330\ day^{-1}$	$\varepsilon_{12} = r_3 = .224\ day^{-1}$	$\varepsilon_{13} = r_4 = .216\ day^{-1}$	$\varepsilon_{14} = r_5 = .216\ day^{-1}$	$\varepsilon_{15} = r_6 = .250\ day^{-1}$
t_{11}				.1512		
t_{12}	.1918			.3310		.1437
t_{13}	.4362		.1317	.7529	.1314	.3269
t_{24}	.1877			.6125	.1289	.3384
t_{31}				.1829		.1079
t_{32}				.1624		
t_{33}	.1474			.4463		.2633
t_{42}	.1333			.3461		
t_{64}	.4125	-.1296		.4735		.4587

Table A.4

Variable	Parameter	
	$\varepsilon_{17} = t_2 = .52\ day$	$\varepsilon_{21} = t_6 = .95\ day$
t_{64}	-.1199	-.7700

Table A.5

Variable	Parameter					
	$\varepsilon_{22} = F_1 =$ .7 fraction	$\varepsilon_{23} = F_2 =$ .7 fraction	$\varepsilon_{24} = F_3 =$ .7 fraction	$\varepsilon_{25} = F_4 =$ .7 fraction	$\varepsilon_{26} = F_5 =$ .7 fraction	$\varepsilon_{27} = F_6 =$ .7 fraction
t_{12}				-.1789		-.1401
t_{13}				-.4069		-.3186
t_{24}		-.1077		-.4389		-.3437
t_{31}				-.1413		-.1106
t_{32}				-.1255		
t_{33}				-.3488		-.2700
t_{42}				-.3108		-.2442
t_{64}	-.1731	-.2368	-.1823	-.9648	-.1962	-.7557

Table A.6

Variable	Parameter	
	$\varepsilon_{31} = Q_4 =$ 3.28 day^{-1}	$\varepsilon_{33} = Q_6 =$ 3.28 day^{-1}
t_{64}	.1998	.1823

Table A.7

Variable	Parameter	
	$\varepsilon_{37} = K_{s_4} =$ 2 $gm/m^2/day$	$\varepsilon_{39} = K_{s_6} =$ 2 $gm/m^2/day$
t_{13}	-.1424	-.1115
t_{24}	-.1536	-.1203
t_{33}	-.1206	
t_{64}	-.3377	-.2645

Table A.8

Variable	Parameter	
	$\varepsilon_{43} = R_{w_4} =$ 3.38 meters	$\varepsilon_{45} = R_{w_6} =$ 2.90 meters
t_{64}	.1998	.1823

Table A.9[a]

Variable	Parameter
	$\varepsilon_{63} = S_6 =$ 2.18 mg/ℓ
t_{11}	.1339
t_{12}	.2930
t_{13}^{+}	.6663
t_{24}^{+}	.7188
t_{31}^{+}	.2314
t_{32}^{+}	.2055
t_{33}	.5646
t_{42}	.5107
t_{64}^{+}	1.5802

[a]Components of t_{ij} marked with a plus sign (+) extrapolate to unity as a result of a one unit increase in the dissolved oxygen deficit S_6 in reach 6.

Table A.10

Variable	Parameter	
	$\varepsilon_{66} = \bar{P}_{2-6} =$ .85 fraction	$\varepsilon_{67} = \underline{PP}_4 =$ .2 fraction
t_{12}	.1612	
t_{13}	.3666	
t_{21}	-1.2210	
t_{22}	-1.0990	
t_{23}	-.9111	
t_{24}	2.5797	
t_{31}	.1264	
t_{32}	.1122	
t_{33}	.3085	
t_{41}		-.9994
t_{42}	.2739	.3040
t_{61}	-1.2212	
t_{62}	-1.0988	
t_{63}	-.9112	
t_{64}	5.6704	

APPENDIX 3 Input Data and Calculated u_{ij} for the Problem P_1

List of Input Data[a]

Parameter	Reach 1	Reach 2	Reach 3	Reach 4	Reach 5	Reach 6
t	4.0000	0.5200	0.7000	2.6200	0.3700	0.9500
r	0.0800	0.3300	0.2240	0.2160	0.2160	0.2500
K	0.1260	0.1260	0.1260	0.1260	0.1260	0.1260
V_E	16.0000	32.5000	6.4000	41.5000	48.5000	20.0000
E	148.0000	436.0000	179.0000	38.0000	4.8300	661.0000
F	0.7000	0.7000	0.7000	0.7000	0.7000	0.7000
R_w	4.8800	2.4400	3.4100	3.3800	3.2000	2.9000
Q	3.2800	3.2800	3.2800	3.2800	3.2800	3.2800
S	2.1800	2.1800	2.1800	2.1800	2.1800	2.1800
K_s	2.0000	2.0000	2.0000	2.0000	2.0000	2.0000

$$\left.\begin{array}{ll} V_{R_0} & 1162 \\ L_{b_0} & 1.0 \\ D_{b_0} & 1.0 \end{array}\right\} \text{Input data upstream of Reach 1}$$

[a]Data from [7].

Calculated u_{ij}

i	j = 1	2	3	4	5	6
1	0.7756	0.0000	0.0000	0.0000	0.0000	0.0000
2	0.8544	0.9172	0.0000	0.0000	0.0000	0.0000
3	0.8804	2.0082	0.1029	0.0000	0.0000	0.0000
4	0.8540	4.3310	0.3305	0.4018	0.0000	0.0000
5	0.8433	4.4959	0.3476	0.4337	0.0120	0.0000
6	0.8162	4.8288	0.3827	0.5003	0.0379	1.6400

REFERENCES

1. R. L. Armacost, Sensitivity analysis in parametric nonlinear programming, D.Sc. dissertation, The George Washington University, Washington, D.C., 1976.
2. R. L. Armacost and A. V. Fiacco, Computational experience in sensitivity analysis for nonlinear programming, *Math. Programming 6*(1974), 301-326.
3. R. L. Armacost and A. V. Fiacco, Sensitivity analysis for parametric nonlinear programming using penalty methods, *Proc. Bicentennial Conf. on Math. Programming*, National Bureau of Standards, 29 November - 1 December 1976; *Computers and Mathematical Programming*, NBS Special Publication 502, U.S. Department of Commerce, 261-269, 1978.
4. R. L. Armacost and W. C. Mylander, A guide to a SUMT-Version 4 computer subroutine for implementing sensitivity analysis in nonlinear programming, Technical Paper T-287, Institute for Management Science and Engineering, The George Washington University, 1973.
5. A. Charnes and R. E. Gemmell, A method of solution of some non-linear problems in abatement of stream pollution, Systems Research Memorandum 103, The Technological Institute, College of Arts and Sciences, Northwestern University, 1964.
6. A. Charnes, Personal correspondence, May 1979.
7. J. G. Ecker, A geometric programming model for optimal allocation of stream dissolved oxygen, *Management Sci. 21*(1975), 658-668.
8. A. V. Fiacco, Sensitivity analysis for nonlinear programming using penalty methods, *Math. Programming 10*(1976), 287-311.
9. A. V. Fiacco and A. Ghaemi, Optimal treatment levels of a stream pollution abatement system under three environmental control policies, Part I: Solution and analysis of convex equivalents of Ecker's GP models using SUMT, Technical Paper T-387, Institute for Management Science and Engineering, The George Washington, 1979.
10. A. V. Fiacco and A. Ghaemi, Sensitivity analysis of a nonlinear structural design problem, Technical Paper Serial T-413, Institute for Management Science and Engineering, The George Washington University, 1979.
11. A. V. Fiacco and G. P. McCormick, *Nonlinear Programming: Sequential Unconstrained Minimization Techniques*, New York: Wiley, 1968.
12. W. C. Mylander, R. L. Holmes, and G. P. McCormick, A guide to SUMT-Version 4: The computer program implementing the sequential unconstrained minimization technique for nonlinear programming, Technical Paper RAC-P-63, Research Analysis Corporation, McLean, Virginia, 1971.

Chapter 9 PROBLEM FORMULATIONS AND NUMERICAL ANALYSIS IN INTEGER PROGRAMMING AND COMBINATORIAL OPTIMIZATION*

STEPHEN C. GRAVES and JEREMY F. SHAPIRO / Massachusetts Institute of Technology, Cambridge, Massachusetts

I. INTRODUCTION

Experienced practitioners who use integer programming (IP) and other combinatorial optimization models have often observed that numerical problems to be optimized are sensitive, sometimes extremely so, to the specific problem formulations and data. Unlike linear programming (LP) and nonlinear programming, however, we have not seen the development of a coherent field of numerical analysis in IP and combinatorial optimization. By numerical analysis, we mean techniques for analyzing problem formulations and data that reveal the stability, or predictability, in optimizing these problems, and the expected degree of difficulty in doing so. Three apparent reasons for the lack of development are:

1. Issues of numerical analysis in IP and combinatorial optimization are intermingled with the artistry of modeling. Thus, unlike LP and nonlinear programming, these issues cannot be related mainly to specific

*Research supported in part by National Science Foundation Grant MSC-77-24654.

numerical problems that have already been generated by the practitioner.

2. IP and combinatorial optimization problems possess a wide range of special structures that can sometimes be exploited by special purpose algorithmic methods. The relative merits of general versus special purpose approaches remains an open question, but the ambiguity has inhibited the development of general purpose numerical analytic methods.
3. Many IP and combinatorial optimization problems can be very difficult to optimize exactly and sometimes even approximately. Systematic procedures for problem formulation and numerical analysis need to be related to approximate as well as exact methods, and the approximate methods are still under development.

Our purpose in this paper is to present a broad sampling of the issues in IP and combinatorial optimization problem formulation, and related questions of numerical analysis. First, primarily through illustrative examples, we discuss the importance of problem formulation. Second, we present briefly some formalisms that can facilitate both our understanding of the art of formulation, and our ability to perform numerical analysis. Finally, we suggest some areas of future research. We believe there is considerable room for the design and implementation of new numerical procedures for the practical solution of IP and combinatorial optimization problems that would greatly expand their usefulness.

II. PROBLEM FORMULATIONS

Integer decision variables arise naturally in many applications, such as airline crew scheduling or investment problems where the items to be selected are expensive and indivisible. Integer variables are also used to model logical conditions such as the imposition of fixed charges. Fixed charge problems are among the most difficult IP problems to optimize, in large part because of the awkwardness of expressing logical conditions by inequalities.

Consider, for example, the problem

$$\min \sum_{j=1}^{n} c_j x_j + \sum_{k=1}^{K} f_k y_k$$

$$\begin{aligned}\text{s.t. } & \sum_{j=1}^{n} a_{ij}x_j \geqq b_i && \text{for } i = 1,\dots,m \\ & \sum_{j\varepsilon J_k} x_j - |J_k|y_k \leqq 0 && \text{for } k = 1,\dots,K \\ & x_j = 0 \text{ or } 1,\ y_k = 0 \text{ or } 1 \end{aligned} \tag{1}$$

where each J_k is an arbitrary subset of $\{1,\dots,n\}$ and $|J_k|$ denotes its size. The quantity f_k is the fixed charge associated with using the set of variables x_j for $j \varepsilon J_k$. The ordinary LP relaxation of (1) is the problem that results if we let the x_j and the y_k take on any values in the range zero to one.

The fixed charge problem (1), and others similar to it, is difficult to solve because the LP relaxations tend to be highly fractional. These LP's are the primary tool used in branch and bound, or other methods, for solving (1); hence there is a tendency for the branch and bound searches to be extensive. For example, suppose we let $\tilde{x}_j$ denote feasible LP or IP values for the x_j variables. If $f_k > 0$ for all k, then the corresponding y_k values for all k are given by

$$\tilde{y}_k = \frac{\sum_{j\varepsilon J_k} \tilde{x}_j}{|J_k|}$$

which in general are fractional numbers.

A number of researchers have observed that problem (1) is easier to solve if the fixed charge constraints are rewritten in the equivalent integer form as

$$x_j - y_k \leqq 0 \qquad \text{for all } j \varepsilon J_k$$

Note that we will have $y_k = \max_{j\varepsilon J_k} \{x_j\}$ if $f_k > 0$, implying $y_k = 1$ if any $x_j = 1$ for $j \varepsilon J_k$. In other words, we can expect the ordinary LP relaxation of the reformulation to be tighter in the sense that it produces solutions with fewer fractions. The reformulation has the effect of *increasing* the number of rows of the problem, but the improved formulation more than justifies the increase. In one application made by one of the authors, a difficult 80-row fixed charge problem was reformulated as indicated as a 200-row problem for which an optimal solution was easily computed. H. P. Williams [30,31] has made a thorough study of fixed charge problem reformulations arising in a variety of applications. Spielberg [29] discusses a variety of approaches to the formulation and analysis of logical inequalities in enumerative methods for integer programming; he also gives an extensive list of references. We will return again to the fixed charge problem in the next section when we discuss algorithmic methods.

Similar observations on the importance of obtaining tight LP formulations have been made for plant location problems by Spielberg [28] and Davis and Ray [6], for a distribution system design problem by Geoffrion and Graves [11] and for a production allocation and distribution problem by Mairs, Wakefield, Johnson, and Spielberg [20]. In all cases, the tightness of the LP relaxation was improved by the careful choice of model representation of the logical relationships.

Another class of IP reformulation "tricks" that has proven successful are procedures for reducing coefficients. For example, consider the inequality

$$114x_1 + 127x_2 + 184x_3 \leq 196$$

to be satisfied by integer values for the variables. The difficulty with this inequality arises again from the nature of LP relaxations to the IP problem with this constraint. If the constraint is binding, then we will surely have a fractional LP solution since no combination of the numbers 114, 127, and 184 will equal 196. However, an equivalent integer inequality is

$$x_1 + x_2 + x_3 \leq 1$$

The inequality with smaller coefficients is less likely to produce fractional solutions in the LP relaxations.

In general, we will not be able to achieve an equivalent representation of an inequality in integer variables with coefficients reduced to the extreme degree of the above example. Gorry, Shapiro, and Wolsey [16] give the following general rule: Any nonnegative solution satisfying

$$\sum_{j=1}^{n} a_{ij}x_j \leq b_i \tag{2}$$

will also satisfy

$$\sum_{j=1}^{n} \left[\frac{a_{ij}}{\lambda_i}\right] x_j \leqq \left[\frac{b_i}{\lambda_i}\right] \tag{3}$$

where λ_i is any positive number and [a] denotes the largest integer smaller than a. In other words, the inequality (3) is a relaxation of the inequality (2). This is a useful property because a relaxed IP problem that is easier to solve can be used to provide lower bounds in a branch and bound scheme for solving the original IP problem from which the relaxation is derived. Of course, the critical issue is the strength of the lower bounds produced by the relaxation. Gorry, et al. [16] report on some computational experience with relaxations of this type. Bradley, Hammer, and Wolsey [4] give more powerful coefficient reduction methods, but ones that can require significant computational effort.

There is a wide variety of other tricks that can be used to restructure IP problems to make them easier to solve. Some are very simple; for example, using a budget constraint $8x_1 + 10x_2 + 15x_3 \leqq 29$ to deduce that the integer variables x_1, x_2, x_3 have upper bounds of 3, 2, and 1, respectively. Tight upper bounds on integer variables can be very important in reducing the search time required by branch and bound. Krabek [19] reports good experience in solving IP and MIP problems that have been automatically reformulated and simplified by tricks such as these. In the following section, we attempt to demonstrate that there is some underlying theory for understanding and integrating many of these tricks.

A central issue in IP and combinatorial optimization problem formulation is the fact that most combinatorial optimization problems can be represented as IP problems, but sometimes these formulations do not permit special structures to be exploited. In this regard, an important class of problems are those that can be represented, perhaps by suitable transformations, as pure and generalized network optimization problems. For these problems, simplex-like network optimization algorithms will produce integer solutions, often in a very efficient manner.

As an example of this, consider the shift scheduling problem. Suppose each 24-hour day is broken into six four-hour periods, where the minimum staffing requirements for the ith period are given as r_i, $i = 1,\ldots,6$. The problem is to determine the minimum work force to satisfy these requirements where each worker works a continuous two-period (eight-hour) shift. Letting x_i be the number of people whose shift starts at period 1, the problem can be stated as

$$\min \sum_{i=1}^{6} x_i$$

$$\text{s.t. } x_1 + \qquad\qquad x_6 \geq r_1$$

$$x_1 + x_2 \qquad\qquad \geq r_2$$

$$\ddots$$

$$x_5 + x_6 \geq r_6$$

$$x_1 \geq 0, \text{ integer}$$

Figure 1 gives a reformulation of this problem as a network flow problem, where w_i is the actual staffing level for period i. Here the arc flows w_i have a lower bound of r_i, while the arc flows x_i have a unit cost of one.

Since many IP and combinatorial optimization problems can be transformed into network optimization problems if one is willing to greatly expand problem size, an open empirical question is the extent of the useful class of such problems. Glover and Mulvey [14] give details about these transformations. Additional discussion about the class of applications that can be formulated and solved as network optimization problems is given by Glover, Hultz, and Klingman [13].

In addition to imbedded network optimization problems, there are many other exploitable special structures that can arise in IP and combinatorial optimization problems. However, it sometimes takes considerable insight to identify the special structures. A prime example of this is the traveling salesman problem and related vehicle routing problems. Held and Karp [17] show that a particular IP formulation of the traveling salesman problem can be recast as a simple graph optimization problem with a small number of side constraints. This reformulation led to great improvements in computation (Held and Karp [18]). An alternative conceptualization of this problem is given by Miliotis [21], who reports good computational experience with an IP formulation of the problem that is built up iteratively in the manner of the IP cutting plane method. A third approach is that of Picard and Queyranne [24], who formulate the time-dependent traveling salesman problem as a shortest path problem on a multipartite graph, supplemented by a set of side constraints.

For the vehicle routing problem, Fisher and Jaikumar [8] formulate the problem such that a natural decomposition arises in which a generalized assignment subproblem with side constraints is solved iteratively with a series of traveling salesman problems. The generalized assignment

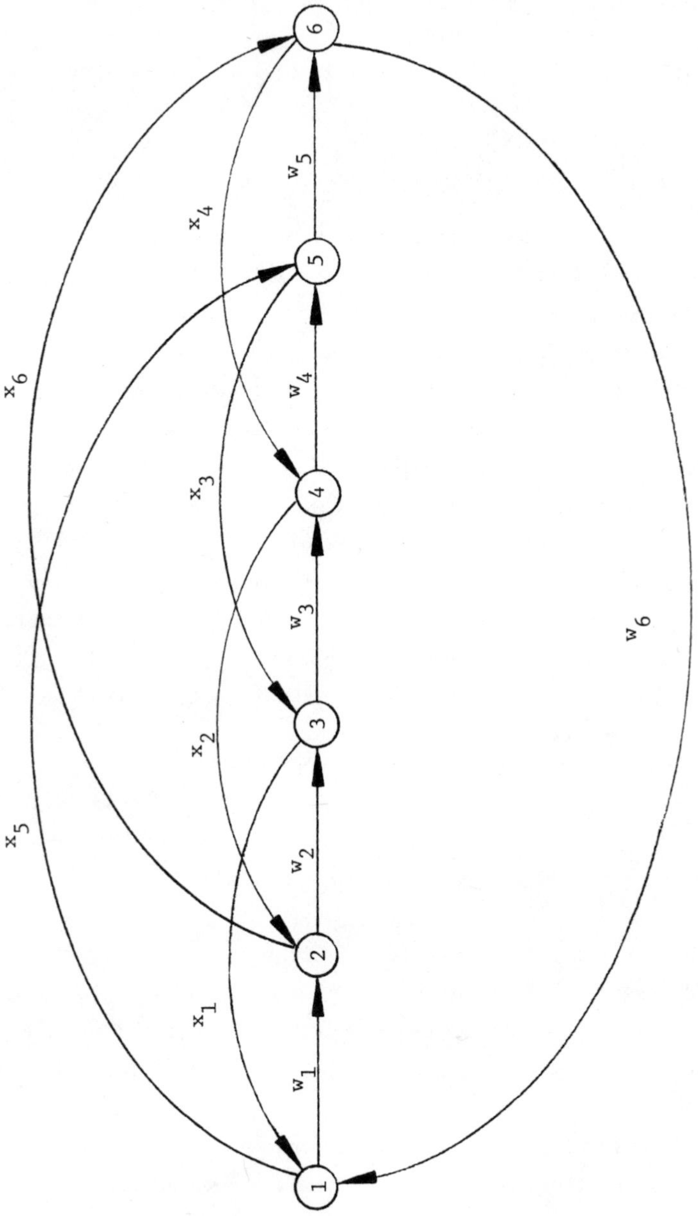

Figure 1

problem divides the cities amongst the vehicles, while each traveling salesman problem generates a route for a specific vehicle. Gavish and Graves [9] give new formulations for the traveling salesman problem and for a variety of related transportation routing problems. These formulations suggest several approaches, which exploit the underlying network structure.

III. ALGORITHMIC METHODS AND NUMERICAL ANALYSIS

We discussed in the previous section how a great deal of artistry is required to select efficient IP and combinatorial optimization problem formulations, and to apply the correct tricks to improve these formulations. Although it may never be possible to develop formalisms to automatically select efficient formulations for all problems, there remains considerable room for developing further the relevant numerical analytic techniques. We discuss how current and future research efforts in three important areas should facilitate these developments: Lagrangian techniques, elementary number theory, and approximation methods.

It is not within the intended scope of this paper to give an extensive survey of Lagrangian techniques applied to IP and combinatorial optimization problems (see, for example, Shapiro [26]). Instead we will review briefly their application to the family of (primal) IP problems, and relate them to the formulation issues discussed in the previous section. The family of IP problems is

$$\begin{aligned} v(y) = \min\ & cx \\ \text{s.t. } & Ax \leqq y \\ & x \in X \end{aligned} \qquad P(y)$$

where A is an $m \times n$ matrix of integers, y is an $m \times 1$ vector of integers, and X is a discrete set in R^n in which the variables must lie. The function v(y) is called the *integer programming perturbation function*. We may be interested in v defined either for a specific or for a family of right-hand sides y. The partition in P(y) is intended to separate the easy constraints $x \in X$ from the difficult ones $Ax \leqq y$. We will discuss below the partition at greater length after we have discussed briefly the Lagrangian methods.

There is an entire family of Lagrangian functions and related dual problems that can be derived from P(y). The simplest is defined for $u \geqq 0$ as

$$L^0(u;y) = -uy + \min_{x \in X}(c + uA)x \tag{4}$$

The value $L^0(u;y)$ is a lower bound on $v(y)$, and the greatest lower bound is found by solving the dual problem

$$\begin{aligned} w(y) &= \max L^0(u;y) \\ \text{s.t. } u &\geqq 0 \end{aligned} \qquad \text{D(y)}$$

In general, $w(y) \leqq v(y)$ and if $w(y) < v(y)$, we say there is a *duality gap* between P(y) and D(y). For fixed $y = \bar{y}$, the lower bounds $L^0(u;\bar{y})$ can be used to fathom subproblems in a branch and bound scheme to solve $P(\bar{y})$ (see Shapiro [25,26]). Moreover, the Lagrangian sometimes provides optimal solutions to subproblems by appeal to the following global optimality conditions. These conditions also provide the rationale for selecting the m-vector of dual variables.

GLOBAL OPTIMALITY CONDITIONS (VERSION ONE). For a given primal IP problem $P(\bar{y})$, the solutions $\bar{x} \in X$ and $\bar{u} \geqq 0$ satisfy the global optimality conditions if

i. $L(\bar{u};\bar{y}) = -\bar{u}\bar{y} + (c + \bar{u}A)\bar{x}$
ii. $\bar{u}(A\bar{x} - \bar{y}) = 0$
iii. $A\bar{x} \leqq \bar{y}$

The following theorem establishes that these conditions provide (globally) optimal conditions to $P(\bar{y})$.

THEOREM 1 If $\bar{x} \in X$, $\bar{u} \geqq 0$ satisfy the global optimality conditions for the primal problem $P(\bar{y})$, then $\bar{x}$ is optimal in $P(\bar{y})$ and $\bar{u}$ is optimal in the dual problem $D(\bar{y})$. Moreover, $v(\bar{y}) = w(\bar{y})$.

PROOF. See Shapiro [25,26].

The global optimality conditions are sufficient but not necessary for a given problem $P(\bar{y})$ in the sense that there may not be any $\bar{u} \geqq 0$ such that an optimal solution $\bar{x}$ will be identified as optimal by the conditions. This is the case when there is a duality gap between $P(\bar{y})$ and $D(\bar{y})$. We will discuss below how duality gaps can be resolved, and how the resolution is related to the formulation of $P(\bar{y})$. As we shall see, methods for resolving duality gaps are derived from representation of the dual problem $D(\bar{y})$ as large scale LP problems. These LP problems are convexified relaxations of the primal problem $P(\bar{y})$, which led Geoffrion [11] to refer to the dual problems as *Lagrangian relaxations*.

The implicit assumption in our definition and proposed use of the Lagrangian $L^0(u;y)$ is that it can be easily computed, relative to the computation of an optimal solution to $P(y)$. The nature of the set X determines whether or not this assumption is valid. In some cases, the implicit constraints $x \in X$ contain the logical ones not involving data that may change or be parameterized; e.g., fixed charge constraints. Moreover, these constraints need not be stated as inequalities if they can be handled directly as logical conditions in the Lagrangian calculation. By contrast, the constraints $Ax \leqq y$ might refer to scarce resources to be consumed or demand to be satisfied and do involve data that may change or be parameterized. In other cases, the constraints $x \in X$ may refer to network optimization substructures that can be solved very efficiently by special algorithms.

Thus, the artistry of IP problem formulation and analysis reduces in part to the scientific question of how best to partition the constraint set into easy and difficult subsets. As an illustration, consider the following formulation of the traveling salesman problem from Gavish and Graves [9]:

$$\min \quad Z = \sum_{i=1}^{n} \sum_{j=1}^{n} c_{ij} x_{ij} \tag{5a}$$

$$\text{s.t.} \quad \sum_{i=1}^{n} x_{ij} = 1 \qquad j = 1,2,\ldots,n \tag{5b}$$

$$\sum_{j=1}^{n} x_{ij} = 1 \qquad i = 1,2,\ldots,n \tag{5c}$$

$$\sum_{\substack{j=1 \\ j \neq i}}^{n} y_{ji} - \sum_{\substack{j=2 \\ j \neq i}}^{n} y_{ij} = 1 \qquad i = 2,\ldots,n \tag{5d}$$

$$y_{ij} \leqq (n - 1)x_{ij} \qquad i = 1,\ldots,n;\ j = 2,\ldots,n;\ i \neq j \tag{5e}$$

$$x_{ij} = 0,1;\ y_{ij} \geqq 0 \tag{5f}$$

Here, x_{ij} denotes the inclusion of the arc connecting node i to j in the Hamiltonian circuit, while y_{ij} may be thought of as the flow along that arc. Three possible Lagrangian relaxations are suggested by this formulation. First, by dualizing on the constraint set (5b) or (5c), we obtain the relaxation given by Held and Karp [17,18], a minimal cost one-tree problem. An alternative relaxation is found by dualizing on the constraint set (5d); here, the resulting Lagrangian can be seen to be an assignment problem. Finally, by dualizing on the forcing constraints in (5e), the Lagrangian separates into an assignment problem in the $\{x_{ij}\}$ variables and a minimum cost network flow in the $\{y_{ij}\}$ variables. The best Lagrangian depends upon the particular problem specification.

Returning to the general IP problem P(y), we can see that some formulation tricks can be interpreted as manipulation of the set X to make it easier to optimize over; e.g., eliminating nonbinding logical constraints, tightening upper bounds on the variables, etc. In addition, the form $L^0(u;y)$ suggests that we may be able to ignore the magnitudes of the coefficients in the system $Ax \leqq y$ since small round-off errors would tend to cancel and could be at least partially accounted for by corresponding variations in u. The validity of this observation is an open research question. As we shall see, the magnitude and accuracy of the coefficients in $Ax \leqq y$ has an effect, probably an important one, on whether or not there is a duality gap between P(y) and D(y).

Resolution of duality gaps for P(y) is achieved by the application of elementary number theory to strengthen the Lagrangian. The number theoretic procedures are usefully formalized by Abelian group theory. First, we need to reformulate the primal problem as

$$\begin{aligned} v(y) &= \min cx \\ \text{s.t. } Ax + Is &= y \\ x \in X,\ s &\geqq 0 \text{ and integer} \end{aligned} \qquad P(y)$$

Let Z^m denote the Abelian group consisting of integer m-vectors under ordinary addition. Let G denote a finite Abelian group and let ϕ denote a homomorphism mapping Z^m onto G. For the moment, we ignore the rationale for G and ϕ. Methods for their selection will be clear after we show how they are used.

The homomorphism ϕ is used to aggregate the linear system $Ax + Is = y$ defined over the infinite Abelian group Z^m to a group equation defined over the finite Abelian group G. The group equation is added to the Lagrangian

$$L^{\phi}(u;y) = -uy + \min\{(c + uA)x + us\} \tag{6a}$$

$$\text{s.t.} \sum_{j=1}^{n} \phi(a_j)x_j + \sum_{i=1}^{m} \phi(e_i)s_i = \phi(y) \tag{6b}$$

$$x \in X,\ s \geqq 0 \text{ and integer} \tag{6c}$$

where e_i = ith unit vector in R^m, a_j = jth column of A.

Although P(y) has been converted to an equality problem, the vectors u are still constrained to be nonnegative because $L^{\phi}(u;y) = -\infty$ for all other dual vectors. The new dual problem is

$$w^{\phi}(y) = \max L^{\phi}(u;y) \qquad D^{\phi}(y)$$
$$\text{s.t. } u \geqq 0$$

We define for all $\beta \in G$ the sets

$$H(\beta) = \{(x,s) \mid (x,s) \text{ satisfy (6b) and (6c) with group right-hand side } \beta \text{ in (6b)}\}$$

Similarly, we define the function

$$Z^{\phi}(u;\beta) = \min\{(c + uA)x + us\} \quad \text{s.t. } (x,s) \in H \tag{7}$$

For any $\beta \in G$ and any $y \in Z^m$ such that $\phi(y) = \beta$, problems (6) and (7) are connected by

$$L^{\phi}(u;y) = -uy + Z^{\phi}(u;\beta)$$

Problem (7) is a group optimization problem with side constraints determined by the set X. The size of G largely determines the relative computational effort to solve (6). If X simply constrains the x_j to be zero-one, or nonnegative integer, then it can be efficiently solved for all $\beta \in$ G for groups of orders up to 5,000 or more (see Glover [12]; Gorry, Northup, and Shapiro [15]; Shapiro [26]). If X contains fixed charge constraints defined over nonoverlapping sets, efficient computation is still possible (Northup and Sempolinski [22]). More generally, the effect of various side constraints in group optimization problems has not yet been fully explored (see also Denardo and Fox [7]). Nevertheless, this is a future research direction of significant importance and promise.

The first version of global optimality conditions can now be adapted to the new Lagrangian.

GLOBAL OPTIMALITY CONDITIONS (VERSION TWO). For a given primal problem $P(\bar{y})$, the solutions $(\bar{x},\bar{s}) \in H$ and $\bar{u} \geqq 0$ satisfy the global optimality conditions if

i. $L^{\phi}(\bar{u};\bar{y}) = -\bar{u}\bar{y} + Z^{\phi}(\bar{u};\ \phi(\bar{y}))$

ii. $A\bar{x} + I\bar{s} = \bar{y}$

The complementary slackness condition of Version One has been omitted because the inequalities were replaced by equations. As before, $(\bar{x},\bar{s})$ is optimal in $P(\bar{y})$ and $\bar{u}$ is optimal in $D^{\phi}(\bar{y})$ if the global optimality conditions hold.

With this background, we can make some observations about how the data of an IP problem affect its optimization. We define the *degree of difficulty* of the IP problem $P(y)$ as the size of the smallest group G for which there is a homomorphism ϕ mapping Z^m onto G such that the global optimality conditions hold relative to some u that is optimal in the induced dual problem. A degree of difficulty equal to one for some $P(y)$ means that the simple Lagrangian L^0 defined in (4) will yield an optimal solution to $P(y)$ via the first version of the global optimality conditions. Primal problems with small degrees of difficulty are not much more difficult to solve. These constructs permit us to study analytically the sensitivity of an IP problem to the data. For instance, there can be a great difference between the degree of difficulty of $P(y)$ and $P(y - e)$. Shapiro [27] presents results characterizing the degree of difficulty concept.

A more flexible approach to IP problem solving is achieved if we view the constraints $Ax \leqq y$ as somewhat soft, in that we may tolerate some slight violations. If this is the case, we may be able to compute much more easily an optimal solution to $P(\tilde{y})$ for $\tilde{y}$ near y because the degree of difficulty of $P(\tilde{y})$ is much less than $P(y)$. Moreover, each solution of the group optimization problem (7) for all $\beta \in G$ yields optimal solutions to $|G|$ primal IP problems. Letting $x(u;\beta)$ and $s(u;\beta)$ denote an optimal solution to (7) with group right-hand side β, we can easily show by appeal to the second version of the global optimality conditions that this solution is optimal in $P(Ax(u) + Is(u))$. This is the principle of *inverse optimization* that has been studied in detail for the capacitated plant location problem by Bitran, Sempolinski, and Shapiro [3].

For studying the resolution of duality gaps, we find it convenient to give an LP representation of the dual problem $D^{\phi}(\bar{y})$. Letting (x^t,s^t), for

$t = 1,\ldots,T$, denote the solutions in the set $H(\phi(\bar{y}))$, the dual problem $D^{\phi}(\bar{y})$ can be re-expressed as the large scale LP

$$\begin{aligned} w^{\phi}(y) &= \min \sum_{t=1}^{T} (cx^{t})\lambda_t \\ \text{s.t.} &\sum_{t=1}^{T} (Ax^{t} + Is^{t})\lambda_t = \bar{y} \\ &\sum_{t=1}^{T} \lambda_t = 1 \\ &\lambda_t \geqq 0 \end{aligned} \tag{8}$$

The following theorem characterizes at what point duality gaps occur and provides the starting point for resolving them.

THEOREM 2 Let $\bar{\lambda}_t > 0$ for $t = 1,\ldots,K$, $\bar{\lambda}_t = 0$ for $t = K + 1,\ldots,T$, denote an optimal solution found by the simplex method to the dual problem $D^{\phi}(\bar{y})$ in the form (8). If $K = 1$, then the solution (x^{1},s^{1}) is optimal in $P(\bar{y})$. If $K \geqq 2$, then the solutions (x^{t},s^{t}) for $t = 1,\ldots,K$ are infeasible in $P(\bar{y})$.

PROOF. See Bell and Shapiro [1].

In the latter case of Theorem 2, the column vectors $Ax^{1} + Is^{1},\ldots, Ax^{K} + Is^{K}$ are used to construct a super group G' of G and a homomorphism ϕ' mapping Z^{m} onto G' such that $Z^{\phi'}(u;\phi(\bar{y})) \geqq Z^{\phi}(u;\phi(\bar{y}))$ for all $u \geqq 0$. Specifically, the solutions (x^{t},s^{t}) for $t = 1,\ldots,K$ are infeasible in (7) for the new group equation. Thus, the new dual problem $D^{\phi'}(\bar{y})$ is strictly stronger than its immediate predecessor $D^{\phi}(\bar{y})$. The procedure is repeated until a dual problem is obtained that provides an optimal solution to the primal problem.

Following the construction of Bell and Shapiro [1], the size of the group G' depends on the size of G and the magnitude of the coefficients of the columns $Ax^{t} + Is^{t}$ for $t = 1,\ldots,K$. Clearly, the scale selected for the constraints $Ax \leqq y$ plays an important role in determining the size of the group encountered and the degree of difficulty of the IP problems of interest. It is far better to measure resources y_i in tons rather than ounces, depending, of course, on the nature of the application. Recently Bell [2] has derived some new procedures that permit the construction of a super group G' satisfying in many cases $|G'| = 2|G|$ or $|G'| = 3|G|$, regardless of the magnitude of the coefficients. Further theoretical and empirical research is needed to understand the importance of scaling in IP optimization.

Heuristics and approximation methods are the third research area that has an important bearing on understanding and achieving efficient IP problem formulations. The heuristic methods can be viewed as working backwards from IP and combinatorial problems with favorable problem structures towards more complex problems to extend the applicability of efficient solution methods. For example, Cornuejols, Fisher, and Nemhauser [5] have derived a "greedy" heuristic similar to an exact algorithm for the spanning tree problem for a class of uncapacitated location problems. The theoretical efficiency of the heuristic is evaluated using Lagrangian techniques. This approach suggests that, when appropriate, IP and combinatorial models should be selected so that they resemble as much as possible simpler models for which efficient solution methods are known.

The inverse optimization approach mentioned above is an exact approximate method providing optimal solutions to some primal problems that hopefully are in a close neighborhood of a given primal problem. This approach can be usefully combined with heuristics to modify the optimal solutions for the approximate problems to make them feasible in the given problem. More generally, specification by the user of a parametric range of interest for the problem specification might make analysis *easier* if it eliminated a slavish concern for optimizing a specific troublesome problem. A related point is that the branch and bound approach to IP and combinatorial optimization generally allows and even relies upon termination of the search of feasible solutions with a feasible solution whose objective function value is within ε of being optimal, where ε is a prespecified tolerance level.

IV. CONCLUSIONS

We have discussed in this paper how efficient IP and combinatorial optimization is highly dependent on problem formulations and data. We also surveyed a number of seemingly unrelated formulation tricks that have been discovered to help achieve efficient formulations. Finally, we argued that formalisms based on Lagrangian techniques, elementary number theory, and approximation methods provide a scientific basis for understanding the tricks. Moreover, the formalisms suggest a new research area devoted to numerical analysis of IP and combinatorial optimization problems.

The application of principles of efficient problem formulation is best achieved at the problem generation stage. One example is the cited approach of Krabek [19], who developed automatic procedures to tighten up MIP problem formulations prior to optimization by a commercial code. Northup and Shapiro [23] report on a general purpose logistics planning system, called LOGS, that generates and optimizes large scale MIP problems from basic decision elements specified by the user. Many of the MIP problem formulation tricks can be brought to bear on the specific MIP model generated from the decision elements.

REFERENCES

1. D. E. Bell, Efficient group cuts for integer programs, *Math. Programming 17*(1979), 176-183.

2. D. E. Bell and J. F. Shapiro, A convergent duality theory for integer programming, *Operations Res. 25*(1977), 419-434.

3. G. R. Bitran, D. E. Sempolinski, and J. F. Shapiro, Inverse optimization: An application to the capacitated plant location problem, Working Paper OR 091-79, MIT Operations Research Center, Cambridge, Massachusetts, 1979.

4. G. Bradley, P. L. Hammer, and L. A. Wolsey, Coefficient reduction for inequalities in 0-1 variables, *Math. Programming* 7(1974), 263-282.

5. G. Cornuejols, M. L. Fisher, and G. L. Nemhauser, Location of bank accounts to optimize float: An analytic study of exact and approximate algorithms, *Management Sci. 23*(1977), 789-810.

6. P. S. Davis and T. L. Ray, A branch-bound algorithm for the capacitated facilities location problem, *Nav. Res. Logist. Quart.* (1969), 331-334.

7. E. V. Denardo and B. L. Fox, Shortest route methods II: Group knapsacks, expanded network, and branch-and-bound, *Operations Res. 27* (1979), 548-566.

8. M. L. Fisher and R. Jaikumar, Solution of large vehicle routing problems, paper presented at TIMS/ORSA Meeting, New York, May 1978.

9. B. Gavish and S. C. Graves, The traveling salesman problem and related problems, Working Paper OR 078-78, MIT Operations Research Center, Cambridge, Massachusetts, 1978.

10. A. M. Geoffrion and G. W. Graves, Multicommodity distribution system design by Benders decomposition, *Management Sci. 20*(1974), 822-844.

11. A. M. Geoffrion, Lagrangian relaxation for integer programming, *Math. Programming Study 2*(1974), 82-114.

12. F. Glover, Integer programming over a finite additive group, *SIAM J. Control 7*(1969), 213-231.

13. F. Glover, J. Hultz, and D. Klingman, Improved computer-based planning techniques: Part II, *Interfaces 9*(1979), 12-20.

14. F. Glover and J. Mulvey, Equivalence of the 0-1 integer programming problem to discrete generalized and pure networks, to appear in *Operations Res.*

15. G. A. Gorry, W. D. Northup, and J. F. Shapiro, Computational experience with a group theoretic integer programming algorithm, *Math. Programming 4*(1973), 171-192.

16. G. A. Gorry, J. F. Shapiro, and L. A. Wolsey, Relaxation methods for pure and mixed integer programming problems, *Management Sci. 18*(1972), 229-239.

17. M. Held and R. M. Karp, The traveling salesman problem and minimum spanning trees, *Operations Res. 18*(1970), 1138-1162.

18. M. Held and R. M. Karp, The traveling salesman problem and minimum spanning trees: Part II, *Math. Programming 1*(1971), 6-25.

19. B. Krabek, Experiments in mixed integer matrix reduction, paper presented at TIMS XXIV International Meeting, Hawaii, June 1979.

20. T. G. Mairs, G. W. Wakefield, E. L. Johnson, and K. Spielberg, On a production allocation and distribution problem, *Management Sci. 24* (1978), 1622-1630.

21. P. Miliotis, Using cutting planes to solve the symmetric traveling salesman problem, *Math. Programming 15*(1978), 177-188.

22. W. D. Northup and D. E. Sempolinski, Fixed-charge integer programming and group optimization problems, in preparation.

23. W. D. Northup and J. F. Shapiro, LOGS: An optimization system for logistics planning, Working Paper OR 089-79, MIT Operations Research Center, Cambridge, Massachusetts, 1979.

24. J. Picard and M. Queyranne, The time-dependent traveling salesman problem and its application to the tardiness problem in one-machine scheduling, *Operations Res. 26*(1978), 86-110.

25. J. F. Shapiro, *Mathematical Programming: Structures and Algorithms*, New York: Wiley, 1979.

26. J. F. Shapiro, A survey of Lagrangian techniques for discrete optimization, in *Annals of Discrete Mathematics 5: Discrete Optimization* (P. L. Hammer, E. L. Johnson, and B. H. Korte, eds.), 113-138, Amsterdam: North-Holland, 1979.

27. J. F. Shapiro, The integer programming perturbation function, in preparation.

28. K. Spielberg, Algorithms for the simple plant-location problem with some side conditions, *Operations Res. 17*(1969), 85-111.

29. K. Spielberg, Enumerative methods in integer programming, in *Annals of Discrete Mathematics 5: Discrete Optimization* (P. L. Hammer, E. L. Johnson, and B. H. Korte, eds.), 139-183, Amsterdam: North-Holland, 1979.

30. H. P. Williams, Experiments in the formulation of integer programming problems, *Math. Programming Study 2*(1974), 180-197.

31. H. P. Williams, The reformulation of two mixed integer programming problems, *Math. Programming 14*(1978), 325-331.

Chapter 10 VALUE PERTURBATION TO REDUCE STORAGE REQUIREMENTS

HARVEY J. GREENBERG and RICHARD P. O'NEILL / Energy Information Administration, U.S. Department of Energy, Washington, D.C.

ABSTRACT

This paper describes a form of purposeful data perturbation in a linear programming model, which pertains to uncertainties in the magnitudes of the matrix coefficients. A problem in value pool construction is described first, then a resolution based on a new concept: "covering lattices." Computer representations of real values, limited by finite precision, is one example of a covering lattice. After presenting the strategy and tactical variations, the effects of resident distortion are analyzed. Several theorems are presented that measure *bias* under a variety of assumptions. For completeness, an appendix is included that contains mathematical proofs and describes considerations pertaining to value pool elimination.

I. INTRODUCTION

Mathematical programming systems, which handle large matrices, use information structures designed to reduce main storage requirements and I/O processing. Pioneering developments exploited *sparsity*--that is, prevalence of zeroes. Most modern systems exploit *supersparsity*--that is, prevalence of replications of nonzero values among the coefficients (first observed by Kalan [6]). The storage reduction comes from the fact that the space to store a floating point value is usually greater than the amount of space required to store an index. For example, on an IBM 370, an index may be stored in two bytes, while a floating point value uses at least four bytes in single precision.

The basic scheme is as follows (see, for example, Greenberg [4]). Every occurrence of a value is recorded by an index that points to a list of distinct values, called a *value pool*. The first phase of input processing, therefore, is to construct the value pool and the index list. Since the structure relevant to this analysis is only the value pool, other design considerations, such as the form of the index list (see [4,5]), will not be discussed.

In current practice, two values are considered to be equal if they have the same bit pattern after normalization. Since the machine representation of values in "floating point" is defined by a mapping from a closed interval of the real line to a finite set, any two numbers are considered equal if their representations are the same (see Forsyth and Moler [2]).

Equality, therefore, is a "compatibility notion" defined by a mapping from the real line to a finite, ordered set of points on the real line--that is, a *lattice*. The matrix coefficient values of a linear programming problem, however, may represent uncertain values, each distributed over a nondegenerate interval. This uncertainty may be contained in original data or accumulated from rounding during computation of the coefficient values. To reflect this uncertainty, a "tolerance" may be specified to indicate that any value in an associated interval may enter in lieu of the specified coefficient, and this interval may be greater than successive machine values--that is, may contain many machine values. In other words, the set of coefficient values to be entered may be mapped to a coarser lattice than the machine's representation of floating point values, if there is some advantage to doing so, without reducing the information content. For example, in many applications, there is little or no meaning past the third or fourth digit of coefficients, and the range of numbers is no more than five orders of magnitude.

First, let us formally define the meaning of compatible values, given two parameters: (1) *relative tolerance* (t_r), and (2) *absolute tolerance* (t_a), where we assume $0 < t_a$ and $0 < t_r < 1$.

DEFINITION Two values, say x and y, are *compatible* if

$$|x - y| \leqq t_r |x + y| + t_a$$

Symbolically, we write $x \sim y$. If two values are not compatible, we say they are *distinct*.

The relevant issues, associated with the construction of the value pool, are discussed in the next section. Emerging from that discussion is a methodology whereby the value pool may contain *no* input value whose entrance is to be recorded. Instead, a list of values is formed where every entrant is compatible with a value in the pool.

Inducing a deviation between entering values and the resident values may be viewed as an instance of *purposeful* data perturbation. This paper considers the *distortion* induced by positive tolerances and presents theorems about resident *bias*. Once the central constructs and fundamental theorems are described, three particular mappings are analyzed: (1) truncation, (2) rounding, and (3) chopping. Finally, some results on solution effects are presented.

II. VALUE POOL CONSTRUCTION

In this section, the construction of the value pool is considered, beginning with a brief review of current practice, where no tolerances are specified. Then, the basic extension is described that deals with tolerance specifications. The important concept introduced is that of a "covering lattice."

A. Current Practice

One algorithm used to process an entering value consists of a linear search over the resident values. If the entrant is already in the value pool, then its location is determined in order to record it in an index (pointer) list. Otherwise, after comparing the entrant with every resident value, the new, distinct value is added to the pool. The effort to construct a value pool using the linear search algorithm just outlined is O(NM), where

N = Number of distinct values
M = Number of nonzero elements

A more efficient scheme, currently practiced, uses a "divide and conquer" strategy (see, for example, Fox [3]). The basic idea is first to partition the value pool into *segments* and assign an entrant to a segment number. (A segment may correspond to a *list* (see Greenberg [4,5]) or a *bucket* (see Kurator and O'Neill [9]), but the exact meaning is not pertinent to our study.) Then, only that segment is searched to locate an equivalent resident value. If the segment does not contain a value that is compatible with the entrant, then the entrant is added to the segment. For a relatively small number of segments (s), this strategy reduces the effort to approximately O(NM/s) (see, for example, Greenberg [5]).

The assignment rule, which is called a *hash function*, is designed to distribute values uniformly over the segments to take full advantage of the strategic reduction of effort to construct the value pool. Such functions separate "nearly equal" values, assigning them different segment numbers (see Knuth [7, p. 508]). When the tolerances are zero, then unequal hash values--segment numbers--must come from distinct values. When, however, the tolerances are positive, then unequal, but compatible, values tend to have different hash values. Thus, the divide and conquer strategy produces a value pool that contains replicated values, counter to the overall objective of frugality with computational resources, namely, time and space.

The question is, "Can the entrance procedure be modified to maintain storage reduction and efficient value pool construction?" The answer is, "Yes," and the next section develops such a modification.

B. Covering Lattices

The problem with using a divide and conquer value pool construction algorithm, while avoiding the entrance of equivalent values to minimize storage, is resolved by emulating machine processing. That is, the idea is first to map the entrant into an a priori set of points, then enter the resulting point. This is formalized by the following.

DEFINITION A *covering lattice*, CL = $\{L(n): n = 0,1,2,\ldots,N\}$ is a collection of increasing values on the interval $[0,V^*]$, such that:

1. $L(0) = 0$
2. $V^* \sim L(N)$
3. For every V in $[0,V^*]$, there exists $L(n) \sim V$.

DEFINITION A covering lattice is *minimal* if no proper subset is a covering lattice.

Two forms of lattices are pertinent to our analysis. The first is the machine's representation of floating point values, and the second is a geometric series. Each has two principal parameters.

DEFINITION A *machine lattice* is the set of points represented by zero and the values generated by the expression:

$$V = r^t \sum_{i=1}^{s} d_i r^{-i}$$

for $t = -e, -e + 1, \ldots, 0, 1, 2, \ldots, E$, $d_1 = 1, 2, \ldots, r - 1$, $d_i = 0, 1, 2, \ldots, r - 1$ for $i = 2, \ldots, s$. The positive, integer-valued parameters, r and s, are the *base* and the *number of significant digits*, respectively.

DEFINITION A *geometric lattice* is the set of points represented by the ex-expression:

$$a(b^n - 1) \qquad \text{for } n = 0, 1, 2, \ldots, N$$

The positive parameters, a and b, are the *reference point* and the *modulus*, respectively.

For a machine lattice, the exponent (t) has range -e to E, so the smallest positive value is r^{-e-1}, and the largest value is $r^E(1 - r^{-s})$. The mantissa may be viewed in the familiar form:

$$\cdot\, d_1\, d_2 \ldots d_s$$

and the requirement that $d_1 \geq 1$ corresponds to a "normalized" form.

The geometric lattice requires that $b > 1$ in order to be a covering lattice. It arises in the formation of a minimal lattice, according to the following.

FUNDAMENTAL RECURRENCE THEOREM Every covering lattice must necessarily satisfy

$$L(n + 1) \leq a + bL(n)$$

where

$$a = 2t_a/(1 - t_r)$$

and

$$b = (1 + t_r)^2/(1 - t_r)^2$$

The recurrence, $L(n + 1) = a + bL(n)$, is sufficient for the lattice to be minimal; in that case

$$L(n) = a(b^n - 1)/(b - 1)$$

The proof of the Fundamental Recurrence Theorem appears in the appendix along with the proofs of other theorems to be presented. Roughly, the rationale is as follows. The first assertion is shown to be compatible with the existence of a point v in the interval (L(n), L(n + 1)) such that $v \sim L(n)$ and $v \sim L(n + 1)$. If no such point exists, then the greatest value compatible with L(n) is strictly less than the least value compatible with L(n + 1). In that case, there is a value not compatible with L(n) or L(n + 1), which violates the definition of a covering lattice. The second assertion stems from constructing the lattice such that v is unique; removal of L(n), for example, implies that no remaining lattice point is equivalent to a value slightly less than v.

The concept of a minimal covering lattice has two deficiencies: (1) there may be a distinguished set of values that must be lattice points, and (2) it is independent of the rule by which entering values are mapped into lattice points. The need to recognize a distinguished set of values arises from a delineation of "value categories." For example, many values are unity (absolute value = +1) to represent an accounting of flows; they must be resident in the value pool (without distortion). That deficiency, therefore, is reconciled by extending the meaning of "covering lattice" and "minimal." We shall not do this here, but we recognize the practical importance of distinguishing certain values, such as integers.

The second property is a deficiency because the notion of "minimal" is considered an "optimality principle." This is true, however, only if storage is the sole consideration. Computation of a lattice point must also be considered, and the geometric formula is nontrivial to process compared to other lattice points, which we shall discuss.

First, let us formalize the notion of mapping entering values into lattice points.

DEFINITION A *lattice map*, f: [0,V*] → CL, satisfies the following conditions:

1. $f(V) \sim V$;
2. $L(n) < V < L(n + 1)$ implies $f(V) = L(n)$ or $f(V) = L(n + 1)$;
3. V in CL implies $f(V) = V$;
4. f is nondecreasing.

DEFINITION Given a covering lattice LC and a lattice map f, we say CL is f-*coarsest* if there exists no proper sublattice for which f is a lattice map.

A sufficient, but not necessary, condition for a covering lattice to be f-coarsest is that it be minimal (since the removal of any lattice point renders the sublattice not covering). The reason an f-coarsest lattice need not be minimal will be illustrated with one lattice map of interest, which we now define.

DEFINITION A lattice map is said to be a *truncation* map (t) if $t(V) \leqq V$ for all V in [0,V*].

We shall show that a truncation map must necessarily be on a covering lattice that is not minimal. The reason, roughly, is that if $V = L(n + 1) - e$, where e is a small positive number, then $t(V) = L(n)$. This implies, therefore, that the lattice must satisfy: $L(n) \sim L(n + 1)$ for all n, in which case the even numbered lattice points comprise a covering lattice, so the complete lattice is not minimal.

TRUNCATION MAP THEOREM If t is a truncation map for CL, then it is necessary that

$$L(n + 1) \leqq a + bL(n) \qquad \text{for } n = 0,1,\ldots,N - 1$$

where

$$a = t_a/(1 - t_r) \qquad \text{and} \qquad b = (1 + t_r)/(1 - t_r)$$

Equality of the above is sufficient for CL to be t-coarsest; in that case,

$$L(n) = a(b^n - 1)/(b - 1)$$

Although a formal proof of the Truncation Map Theorem appears in the appendix, let us explain the rationale. The inequality associated with the first assertion says that it must be true that $L(n) \sim L(n + 1)$. The proof follows from: $t[L(n + 1) - e] = L(n)$ for all $e > 0$, sufficiently small, so we require $L(n + 1) - e \sim L(n)$ for all $e > 0$, sufficiently small. This is true only if $L(n + 1) \sim L(n)$. The sufficiency of the recurrence, $L(n + 1) = a + bL(n)$, to be t-coarsest may be viewed by the property: $L(n)$ is "barely compatible" with $L(n + 1)$--that is, $L(n) \sim L(n + 1)$ but $L(n) \not\sim L(n + 1) + e$ for all $e > 0$. Thus, removal of $L(n)$ would result in $t[L(n + 1) - e] = L(n - 1) \not\sim L(n + 1) - e$ for all $e > 0$, sufficiently small.

Another lattice map of interest is to choose the nearest lattice point in order to minimize average distortion. Formally, we have the following.

DEFINITION A lattice map is said to be a *rounding* map (r) if:

$$|r(V) - V| = \min\{|L(n) - V|: n = 0,1,\ldots,N\}$$

Unlike the truncation map, the rounding map can be defined for *any* covering lattice. Moreover, one lattice that is r-coarsest (besides the geometric lattice given in the Fundamental Recurrence Theorem) is the even numbered lattice points of any t-coarsest lattice. In a sense, therefore, r is coarser than t. This ordering can be formalized, and it is not difficult to define a parametric family of lattice maps by a homotopy from r to t, which are of decreasing coarseness. The point, however, of considering the truncation map is that it is faster to compute than the rounding map. It is with speed of computation in mind that we consider the next mapping, which is derived from the machine representation of normalized, floating point values.

Let r and s be the two parameters of a machine lattice defined earlier, namely, the base and number of significant digits, respectively. In the following definition, we restrict the set of entering values to be machine floating point representations.

DEFINITION A lattice map is said to be a *chopping* map (c) if

$$c(V) = r^t \sum_{i=1}^{s-k} d_i r^{-i}$$

for

$$V = r^t \sum_{i=1}^{s} d_i r^{-i} \qquad \text{where } 1 \le k \le s$$

CHOPPING MAP THEOREM If c is a chopping map with parameters (r,s), and if:

1. $t_a r^{-E+s}/(1 - t_r) \ll 1$
2. $t_r/(1 - t_r) \leq r^{-x}$, x integer
3. $r,s > 1$; and
4. $0 \leq x < s$

then

$$k \leq s - x$$

The Chopping Map Theorem describes a limit (s - x) on how many significant digits can be dropped, yet form a covering (machine) lattice. The chopping map is generally less coarse than a truncation map, in that it requires more points to cover the value interval, [0,V*]. The advantage, however, is the computation of c(V) if we apply the lattice map to the machine's floating point representation. Only a simple logical operation is needed to turn off a specified number of (lowest) bits.

In the next section, we analyze the distortion associated with constructing a value pool by first applying a lattice map to an entering value.

III. RESIDENT DISTORTION

In this section, a probabilistic approach is applied to analyze *distortions*--that is, the differences between original values and their representative lattice points. The mathematical properties derived serve as a foundation to examine the propagation of distortions during the algorithmic process, and finally, to computed solution values.

A. Terms and Concepts

By agreeing to consider unequal values as compatible, the analyst implies that the "true value" is either unknown with certainty or that the relational information it embodies is "fuzzy" due, for example, to the inexact structure of the model. The input values are thus regarded as representative *samples* from an *interval of compatibility* defined by the tolerances--that is, the set of values that are compatible with the entrant. Any value in an entrant's interval of compatibility may be substituted without loss in information content. (As mentioned earlier, some exceptions may be specified, such as unity (1.0), whereupon the covering lattice may be modified to include the exceptions.)

DEFINITION The *distortion* of a value, V, given f(V) = L(n), is the difference

$$D_n(V) = V - L(n)$$

Since the entrant values are no longer known, their distortions are unknown with certainty. Regarding f(V) as "observed," however, we shall consider the *conditional distribution function* of V, given f(V) = L(n), which we denote by

$$F_n(v) = \Pr\{V \leq v \mid f(V) = L(n)\}$$

DEFINITION The *bias* of a lattice point L(n) is the expected distortion

$$B(n) = E\{D_n(V) \mid f(V) = L(n)\}$$

The lattice point is *unbiased* (for a particular distribution function) if B(n) = 0.

The bias has been defined in absolute terms, rather than relative to the value, because it suffices to consider cases where all biases are of the same sign. To begin, let $I_n = \langle l(n-1), l(n) \rangle$ be the interval for which V in I_n implies f(V) = L(n), where exactly one end of I_n is closed (see Appendix 1). Then, the distribution function must satisfy:

$$F_n(v) = 0 \quad \text{if} \quad v < l(n-1)$$

and

$$F_n(v) = 1 \quad \text{if} \quad v > l(n)$$

with equality at exactly one of the two end points. Since I_n contains L(n), it follows that the bias B(n) is signed according to the relation between L(n) and the mean of F_n, say u_n. That is,

$$B(n) = u_n - L(n)$$

implies, for example, $B(n) \geq 0$ for all n if $u_n \geq L(n)$ for all n. This is the case when F_n is uniformly distributed and $[l(n-1) + l(n)]/2 \geq L(n)$, which is true for a class of lattices we now define.

DEFINITION A lattice is (*strongly*) *telescopic* if the first difference is nondecreasing (increasing). That is,

$$\Delta^2 L(n) \geqq (>)\ 0 \qquad \text{for all } n = 2,\ldots,N - 1$$

In particular, it is not difficult to prove that machine lattices are telescopic, and geometric lattices are strongly telescopic.

A companion concept to telescopic lattices is that the midpoint of a lattice interval usually exceeds the lattice point itself, to reflect that there are more values greater than L(n) than less, which map to L(n). This reflects the meaning of "relative tolerance," where absolute distortion may be larger when the value is larger; thus, it is usual that L(n) is less than the mean value mapping to L(n). This is true for the three lattice maps we defined.

DEFINITION A lattice map is said to be *usual* if

$$[l(n - 1) + l(n)]/2 \geqq L(n)$$

BIAS THEOREM If F_n is a distribution over I_n, whose mean is not less than the midpoint of I_n, and the mapping is usual, then the representation bias B(n) of L(n) is nonnegative for all n.

COROLLARY If F_n is uniform and the lattice is (strongly) telescopic, then B(n) is nonnegative (positive) for all n.

The corollary says that for uniformly distributed values, the most common approximation methods for representing real numbers in a machine (namely, a machine lattice) have a nonnegative bias; geometric lattices, such as the t-coarsest and r-coarsest given earlier, have positive bias since they are strongly telescopic.

IV. SOLUTION DISTORTION

In this section, we consider the effect of tolerances on solution values. Our analysis is limited only to costs and right-hand sides in the LP:

$$z = \min\{cx: Ax = b,\ x \geqq 0,\ \sum x_j = v\}$$

The normalization constraint is added to maintain constant sum activity levels; we assume throughout:

$$f(v) = v$$
$$f(A_{ij}) = A_{ij}$$

RIM COMPATIBILITY THEOREM If $f(b) = b$ and $c \geqq 0$, then

$$\frac{z}{v} \sim \frac{\hat{z}}{v}$$

where

$$\hat{z} = \min\{f(c)x\colon Ax = b,\ x \geqq 0,\ \sum x_j = v\}$$

Although the proof is in the appendix, let us explain the rationale. If only costs are perturbed, then the optimal solution to the undistorted problem is still feasible, having the same total activity, and its total cost distortion is bounded by the maximum individual distortions (noting $f(c) \sim c$). When unitizing the cost, namely, z/v, the maximum distortion is also unitized, thus making the original optimal unit cost equivalent to the distorted optimal unit cost. A dual result holds when we assume $f(c) = c$ and permit distortion on the right-hand sides (b). The matrix of technology coefficients (A_{ij}) embodies both computed numbers, where tolerances are relevant, and "logical numbers," which are not perturbed, such as ±1 used to form an accounting structure. A pure transportation problem illustrates a linear program where all values in the body--that is, the A matrix--are logical numbers, and perturbation has little or no meaning. Mathematically, it is well known that the optimum value need not be a continuous function of the body coefficients, so there is no general extension of the Rim Compatibility Theorem.

APPENDIX Proofs of Theorems

We shall prove the theorems presented in the main text. First, it is useful to establish the following.

LEMMA 1 The inverse of a lattice map is an interval, say $f^{-1}(L(n)) = \langle l(n-1), l(n) \rangle$, that contains $L(n)$.

PROOF. If $V1 < V < V2$ and $f(V1) = f(V2) = L(n)$, then monotonicity of f ensures $f(V) = L(n)$, so f is convex--that is, $f^{-1}(L(n))$ is an interval.

Since $f(L(n)) = L(n)$, it follows that $L(n)$ is in the interval of values that map to $L(n)$--that is, $l(n - 1) \leq L(n) \leq l(n)$.

It is also convenient to define I_n as the interval of values compatible with $L(n)$. Since I_n is closed and bounded, we may write: $I_n = [i_n, s_n]$, where

$$i_n = \inf\{V: V \sim L(n)\}$$

and

$$s_n = \sup\{V: V \sim L(n)\}$$

Clearly,

$$i_n \leq l(n - 1) \leq l(n) \leq s_n$$

and

$$s_{n-1} \geq i_n$$

The last inequality follows because otherwise values in (i_n, s_{n-1}) are not covered--that is, are not equivalent to any lattice point.

LEMMA 2 The lattice interval, I_n, satisfies

$$s_n \leq L(n)B + A$$

and

$$i_n \geq L(n)/B - A/B$$

where

$$B = (1 + t_r)/(1 - t_r) \quad \text{and} \quad A = t_a/(1 - t_r)$$

The first part of the Fundamental Recurrence Theorem follows directly from Lemma 2, plus the fact that $A/B \leq A$ since $B \geq 1$. That is,

$$i_{n+1} \leq s_n \leq L(n)B + A$$

and

$$i_{n+1} \geq L(n + 1)/B - A/B$$

implies

$$L(n + 1) \leq L(n)B^2 + A(B + 1)$$

which is the desired result since $b = B^2$ and $a = 2A$. The sufficient condition for CL to be minimal is constructed so that $s_{n-1} = i_n$ for all n. Removal of a lattice point, say L(n), must, therefore, leave values in (s_{n-1}, i_{n+1}) uncovered. (That $s_{n-1} < i_{n+1}$ follows from the fact that $s_{n-1} = i_n < i_{n+1}$ since $t_a > 0$ and $t_r < 1$.) Thus, the Fundamental Recurrence Theorem is proven.

To prove the Truncation Map Theorem, we use the fact that $t^{-1}(L(n)) = (L(n), L(n+1))$, so $l(n) = L(n+1)$ for all n. Thus,

$$L(n+1) = l(n) \leqq s_n \leqq L(n)B + A$$

If equality holds, then removal of L(n) leaves the interval (L(n),L(n+1)) uncovered since $L(n) < V < L(n+1)$ implies $t(V) = L(N-1)$, but $s_{n-1} < i_{n+1}$, so $t(V) \not\sim V$.

The Chopping Map Theorem is now proven. In order for $c(V) \sim V$, it is necessary that:

$$r^t \sum_{i=1}^{s} d_i r^{-i} - r^t \sum_{i=1}^{s-k} d_i r^{-i} \leqq t_r r^t \left(\sum_{i=1}^{s} d_i r^{-i} + \sum_{i=1}^{s-k} d_i r^{-i} \right) + t_a$$

for all t and d. Rearranging terms, this becomes:

$$\sum_{i=s-k+1}^{s} d_i r^{-i}(1 - t_r) \leqq 2t_r \sum_{i=1}^{s-k} d_i r^{-i} + t_a r^{-t}$$

The extreme case, therefore, is for

$d_i = r - 1$ for $i = s - k + 1, \ldots, s$

$d_i = 0$ for $i = 2, \ldots, s - k$

$d_1 = 1$

$t = E$

Substituting these into the necessary inequality, we derive:

$$(r-1) \sum_{i=s-k+1}^{s} r^{-i}(1 - t_r) \leqq 2t_r r^{-1} + t_a r^{-E}$$

The left-hand side simplifies to:

$$-(1 - t_r)(r^{-s} - r^{-s+k}) \leqq 2t_r r^{-1} + t_a r^{-E}$$

By multiplying by r^s, we derive the necessary condition

$$r^k \leqq 1 + 2t_r r^{s-1}/(1 - t_r) + t_a r^{-E+s}/(1 - t_r)$$

By assumption (1), we may ignore the last term, and using assumption (2), k must satisfy:

$$r^k \leqq 1 + 2r^{s-x-1}$$

It remains to be shown that the derived inequality implies the result: $k \leqq s - x$. Equivalently, define $g(k) = 1 + 2r^{s-x-1} - r^k$ and we shall prove $g(k) < 0$ for $k > s - x$. Let p be the positive integer: $p = k + x - s$, so

$$\begin{aligned} g(k) &= 1 + 2r^{s-x-1} - r^{s-x+p} \\ &= 1 + r^{s-x-1}(2 - r^{p+1}) \end{aligned}$$

Since $r \geqq 2$, the last term is negative; in particular,

$$g(k) \leqq 1 + r^{s-x-1}(-2) < 0$$

To prove the Rim Compatibility Theorem, we first bracket $\hat{z} - z$:

$$\sum_j f_j \hat{x}_j - z = \hat{z} - z = \hat{z} - \sum c_j x_j$$

where $f_j = f(c_j)$. Using the definition of compatibility, then rearranging, we obtain:

$$\begin{aligned} \sum f_j \hat{x}_j - z &\geqq \sum \hat{x}_j \left(c_j - t_r(c_j + f_j) - t_a\right) - z \\ &= -t_r \sum \hat{x}_j (c_j + f_j) - t_a v + \sum c_j (\hat{x}_j - x_j) \\ &= -t_r (z + \hat{z}) - t_a v + \sum c_j (\hat{x}_j - x_j)(1 - t_r) \end{aligned}$$

Since $t_r < 1$ and since $\sum c_j (\hat{x}_j - x_j) \geqq 0$,

$$\hat{z} - z \geqq -t_r(z + \hat{z}) - t_a v$$

Again, by using the definition of compatibility and rearranging,

$$\begin{aligned} \hat{z} - \sum c_j x_j &\leqq \hat{z} - \sum x_j \left(f_j - t_r(c_j + f_j) - t_a\right) \\ &= t_r \sum x_j (c_j + f_j) + t_a v + \sum f_j (\hat{x}_j - x_j) \\ &= t_r (z + \hat{z}) + t_a v + \sum f_j (\hat{x}_j - x_j)(1 - t_r) \end{aligned}$$

Since $t_r < 1$ and since $\sum f_j (\hat{x}_j - x_j) \leqq 0$,

$$\hat{z} - z \leqq t_r(z + \hat{z}) + t_a v$$

Therefore,

$$\left|\frac{\hat{z}}{v} - \frac{z}{v}\right| \leqq t_r\left(\frac{z}{v} + \frac{\hat{z}}{v}\right) + t_a$$

That is,

$$\frac{\hat{z}}{v} \sim \frac{z}{v}$$

Value Pool Elimination

This section presents two assembler code segments that perform an inner product calculation for matricial forms [for example, PI*A(j)]. The first data structure stores the row index and a compact representation of the value (using a base plus displacement concept) in each word:

```
     ROW-INDEX/VALUE-DISPLACEMENT PAIR WORD STRUCTURE
 0         14 15  15 16      18 19                       31
-----------------------------------------------------------
I ROW INDEX I SIGN I EXPONENT I MANTISSA                 I
-----------------------------------------------------------
```

This representation allows exponents from 16**-3 (approximately 10**-4) to 16**4 (approximately 10**5), a 13-bit unnormalized fraction (approximately 4 decimal digits), and the exact representation of integers from -4095 to 4095. One endpoint of the exponent range is arbitrary and can be changed by adjusting the base value. This degree of precision is sufficient for many applications. In EIA's Midterm Market Forecasting Model (3000 by 10000), for example, all values (excluding the RHS and boundrow) are in this range. Furthermore, most matrices with values outside this range are notoriously unstable and often must be rescaled before they become solvable.

The second data structure, which is used in many mathematical programming systems (for example, see Krabek, et al. [8]), uses the row-index/value-index data structure for storage of nonzeroes:

```
        ROW-INDEX/VALUE-INDEX PAIR WORD STRUCTURE
 0                        15 16                          31
-----------------------------------------------------------
I ROW INDEX                 I VALUE INDEX                I
-----------------------------------------------------------
```

Although the first code segment has more instructions, it has only two fetches from main memory (real or virtual) if the program is in cache memory. The second requires three fetches from main memory, each of which may be on a different page. The additional instructions in the first code segment are register-to-register and shift instructions, which are among the fastest in the instruction set.

A set of sufficient conditions for the first loop to be faster than the second is: (1) the program is in cache memory (or a faster store); (2) a store and fetch from cache takes less time than one fetch from main memory; and (3) enough needed entries in the value pool are outside of cache to compensate for the extra instructions. Even if the value pool is entirely in cache, which is unlikely, the first loop is not much slower than the second. On the other hand, if one or two of the value pool entries are not in cache then the first loop can be dramatically faster.

The following IBM 360/370 assembly code segment performs the inner product operation using the row-index/value-displacement pair data structure. The nonzero pairs for each column are assumed to be stored contiguously, starting at FIVP and ending at LIVP.

```
       LE   0,0        CLEAR FP REG 0
       L    6,BASE     BASE IS THE BASE VALUE ADDRESS
       LA   7,FIVP     LOAD THE ADDRESS OF THE FIRST PAIR
       LA   8,4        INCREMENT BY WORDS
       LA   9,LIVP     LOAD THE ADDRESS OF THE LAST PAIR
LOOP   EQU  *          LOOP THROUGH THE NONZEROS
       SR   4,4        CLEAR REG 4
       L    5,0(7)     LOAD THE FIRST PAIR
       SLDL 4,15       PUT THE INDEX (1ST 15 BITS) IN R4
       SLL  4,2        MULTIPLY BY 4 FOR WORD DISPLACEMENT
       LE   2,PI(4)    LOAD THE PI VALUE INTO FP REG 2
       SLDL 4,1        MOVE THE SIGN BIT INTO REG 4
       SRL  5,4        POSITION THE DISP. OF THE FP NUMBER
       AR   5,6        CONSTRUCT THE FP NUMBER
       SRDL 4,1        PUT IN THE SIGN
* THIS SEQUENCE OF INSTRUCTIONS IS NEEDED SINCE THERE ARE
* NO GENERAL REGISTER TO FP REGISTER INSTRUCTIONS FOR THE
* 360/370 SERIES. THE USE OF THE IN-LINE TEMPORARY STORAGE
* LOCATION WILL PROMOTE THE USE OF CACHE MEMORY STORES AND
* FETCHES.
       ST   5,STORE    STORE THE FP NUMBER FOR LATER FETCH
       B    JUMP       JUMP OVER THE TEMPORARY STORAGE WORD
STORE  NOP
JUMP   ME   2,STORE    MULTIPLY AJ(INDEX)*PI(INDEX)
       AER  0,2        ACCUMULATE THE INNER PRODUCT
       BXLE 7,8,LOOP   TEST AND BRANCH TO LOOP
* END OF LOOP. THE INNER PRODUCT IS IN FP REG 0
       DS      OF
BASE   DC      X'76000000'
```

The following IBM 360/370 assembly code segment performs the inner product operation using the row-index/value-index pair data structure. The nonzero pairs for each column are assumed to be stored contiguously, starting at FIVP and ending at LIVP. The instructions that are unnecessary in the second loop are prefixed by **.

```
       LE   0,0          CLEAR FP REG 0
       LA   7,FIVP       LOAD THE ADDRESS OF THE FIRST PAIR
       LA   8,4          INCREMENT BY WORDS
       LA   9,LIVP       LOAD THE ADDRESS OF THE LAST PAIR
LOOP   EQU  *            LOOP THROUGH THE NONZEROS
       SR   4,4          CLEAR REG 4
       L    5,0(7)       LOAD THE FIRST PAIR
       SLDL 4,16         PUT THE INDEX (1ST 16 BITS) IN R4
       SLL  4,2          MULTIPLY BY 4 FOR WORD DISPLACEMENT
       LE   2,PI(4)      LOAD THE PI VALUE INTO FP REG 2
**       SLDL 4,1
       SRL  5,14         POSITION THE INDEX OF THE FP NUMBER
**       AR   5,6
**       SRDL 4,1
**       ST   5,STORE
**       B    JUMP
**STORE NOP
* VP IS THE BASE OF THE VALUE POOL
JUMP   ME   2,VP(5)      MULTIPLY AJ(INDEX)*PI(INDEX)
       AER  0,2          ACCUMULATE THE INNER PRODUCT
       BXLE 7,8,LOOP     TEST AND BRANCH TO LOOP
* END OF LOOP. THE INNER PRODUCT IS IN FP REG 0
```

Acknowledgment

The authors wish to thank James E. Kalan for stimulating discussions and Patricia A. Green for her typing.

REFERENCES

1. I. S. Duff and J. K. Reid, Some design features of a sparse matrix code, *ACM TOMS* 5(1979), 18-35.

2. G. E. Forsyth and C. B. Molder, *Computer Solution of Linear Algebraic Systems*, Englewood Cliffs, NJ: Prentice-Hall, 1967.

3. B. L. Fox, Data structures and computer science techniques in operations research, *Operations Res.* 26(1978), 686-717.

4. H. J. Greenberg, A tutorial on matricial packing, in *Design and Implementation of Optimization Software* (H. J. Greenberg, ed.), 109-142, The Netherlands: Sijthoff and Noordhoff, 1978.

5. H. J. Greenberg, Information structures for matricial forms, EIA Technical Report, forthcoming, 1980.

6. J. E. Kalan, Aspects of large-scale in core linear programming, *Proc. ACM* (1971), 304-313.

7. D. E. Knuth, *The Art of Computer Programming*, Vol. 3, Reading, Mass.: Addison-Wesley, 1973.

8. C. B. Krabek, R. J. Sjoquist, and D. C. Sommer, The Apex systems, past and future, CDC Technical Report, Minneapolis, Minnesota, 1979.

9. W. Kurator and R. P. O'Neill, PERUSE: An interactive system for mathematical programs, EIA Technical Report, forthcoming, 1980.

10. W. Orchard-Hays, *Advanced Linear Programming Computing Techniques*, New York: McGraw-Hill, 1968.

Chapter 11 APPLICATION OF CONTINUATION METHODS IN STABILITY AND OPTIMIZATION PROBLEMS OF ENGINEERING

RAMAN K. MEHRA and ROBERT B. WASHBURN, JR. / Scientific Systems, Inc.,
Cambridge, Massachusetts

ABSTRACT

Continuation methods for solving nonlinear equations numerically are applied to a number of engineering applications. The method is used to compute bifurcation surfaces for the study of transient stability of power systems and the stability of aircraft at high angles of attack. Combined with a shooting method for solving two-point boundary value problems, the continuation algorithm can be used to compute limit cycles and optimal trajectories in some aircraft problems. In addition to conventional continuation methods, a combination of continuation with singular perturbation methods of order reduction has been found to be advantageous in trajectory optimization.

I. INTRODUCTION

For applied mathematicians and engineers who must analyze the difficult nonlinear mathematical models arising in engineering problems, perturbation techniques are the chief tools of the trade. The different types of perturbation techniques are too numerous to list here, but all methods attempt in concept to analyze a mathematical problem $P(x_1)$ dependent on data parameter x_1 in terms of another problem $P(x_0)$ with data parameter x_0 close to x_1. The basic approach is to choose x_0 so that $P(x_0)$ is easy to solve and then use a perturbation method to extrapolate the solution of $P(x_0)$ to the solution of the real problem $P(x_1)$. The mathematical foundation for this approach is the differential and integral calculus, and the extrapolation is usually some type of asymptotic series expansion of $P(x_1)$ in the data perturbation $\delta = x_1 - x_0$. For specific techniques the reader should refer to the recent book of Bender and Orszag [1], which provides an excellent introduction to several different perturbation methods and contains an abundance of examples illustrating their application.

Perturbation methods have proved effective tools of analysis in many problems for nearly two centuries, but these methods suffer one common handicap: the perturbation δ must be small for successful application of the method. When the perturbation is not small, the asymptotic series used for extrapolation may require a large number of terms to achieve reasonable accuracy. In large dimensional, complex, nonlinear problems each successive term of the asymptotic series adds a tremendous computational burden. Two or three terms are often all that one can feasibly calculate. Moreover, it can happen that no number of terms, no matter how large, can provide a reasonably accurate solution. For example, consider a power series expansion with a finite radius of convergence.

In this paper we discuss alternative methods, generally described as *continuation* methods, which do not require that the perturbation δ be small. Conceptually, the continuation methods take an approach similar to that of perturbation methods. The continuation methods try to solve the desired problem $P(x_1)$ in terms of an easy problem $P(x_0)$. However, the data parameter x_0 is not required to be close to x_1; what is required is that it be possible to solve $P(x_\lambda)$ for a continuous family of data parameters x_λ with $0 \leq \lambda \leq 1$. The object of these methods is to continue the solution of $P(x_\lambda)$ from the easy parameter value at $\lambda = 0$ to the desired value at $\lambda = 1$. Mathematically, continuation methods rely on differential calculus, just as perturbation methods do, but in addition, continuation methods utilize some abstract results of differential topology. Computationally, continuation methods depend much more on the modern digital computer than perturbation methods do. The combination of differential calculus, topology, and the digital computer provides an effective tool for studying difficult, nonlinear problems of engineering for which classical perturbation methods are inadequate.

The paper is organized into two main sections. The first describes continuation methods in more detail. We briefly give the main types of continuation, the bifurcation difficulties one encounters in trying to apply the method, and a brief history and survey of current work. In the next section we discuss some stability and optimization problems to which we have applied continuation methods. The stability problems concern the dynamic stability of aircraft and power systems. The optimization problems concern the optimization of aircraft trajectories. In the latter case we also describe the application of *asymptotic continuation*, a combination of continuation and singular perturbation methods. A final section concludes the paper and discusses unsolved problems which have significance for the application of continuation methods.

II. CONTINUATION METHODS

A. Introduction

An inherent limitation of all perturbation methods is that the perturbation approximation is valid and accurate only for sufficiently small values of the perturbation parameter. Unfortunately, it sometimes happens that the interesting parameter values are not sufficiently small. This is particularly true when a perturbation parameter is introduced artificially, more for mathematical convenience than for physically justifiable reasons. Continuation methods are techniques for solving parameterized equations which do not require the parameter to be small in some sense. Thus, the continuation methods tend to give solutions over a larger region of parameter space than perturbation methods.

The basic *continuation* problem is to solve the equation

$$G(x,\lambda) = 0 \tag{1}$$

for the vector x in terms of the real parameter λ. If B is a vector space and G maps $B \times R$ into B, then we wish to find the trajectories $\lambda \to x(\lambda)$ in B which satisfy

$$B(x(\lambda),\lambda) = 0 \tag{2}$$

for all parameter values λ. For example, (1) might be the nonlinear equation for the missing initial values in the two-point boundary value problem (TPBVP) from the necessary conditions for an optimal control problem. In this case, B would be a finite dimensional space, $B = R^n$. On the other hand, one might treat the entire trajectory control and state trajectory as a vector in an infinite dimensional vector space B. In that case, (1) would be a nonlinear operator equation given by the Euler-Lagrange necessary conditions for the trajectory optimization problem.

The basic approach of continuation methods is to solve equation (1) at the values $\lambda = \lambda_0, \lambda_1, \lambda_2, \ldots, \lambda_m$ by first solving it at $\lambda = \lambda_0$ and then at each step i using the solution $x(\lambda_{i-1})$ as the initial guess in a Newton-Raphson solution for $x(\lambda_i)$. This is the basis for the method of Lahaye [2]. Davidenko [3], on the other hand, differentiated the equation (2) with respect to λ to obtain an implicitly defined differential equation

$$\frac{\partial G}{\partial x}(x,\lambda)\frac{\partial x}{\partial \lambda} + \frac{\partial G}{\partial \lambda}(x,\lambda) = 0 \tag{3}$$

for x as a function of λ. Discrete approximations of (3) then give formulas to determine the solution $x(\lambda_i)$ in terms of the solution $x(\lambda_{i-1})$. An efficient continuation algorithm will employ both methods, using Davidenko's method as a prediction step and Lahaye's method as a correction step in a way analogous to predictor-corrector integration algorithms.

To illustrate the basic idea of continuation we discuss the use of Davidenko's method to solve a simple quadratic equation, and we compare this method to a perturbation series expansion. Consider the equation

$$y^2 + y = \lambda \tag{4}$$

for real λ which has the exact solutions

$$y_1(\lambda) = \frac{-1 + \sqrt{1 + 4\lambda}}{2} \tag{5}$$

$$y_2(\lambda) = \frac{-1 - \sqrt{1 + 4\lambda}}{2} \tag{6}$$

Suppose we want the solutions of (4) for λ near 1. The solutions of (4) near $\lambda = 0$ have perturbation series expansions which one can calculate directly from (4). These are

$$y_1(\lambda) = \lambda - \lambda^2 + \ldots \tag{7}$$

$$y_2(\lambda) = -1 - \lambda + \lambda^2 + \ldots \tag{8}$$

The approximations (7),(8) to (5),(6) are good when λ is near 0, but the trouble is that $\lambda = 1$ is not small. In fact, the series converge only for $|\lambda| < 1/4$. For $|\lambda| > 1/4$ the series diverge and one cannot use finitely many terms to approximate the exact answer. Moreover, for λ less than but very close to 1/4, the series converge very slowly, and one must calculate very many terms to obtain an accurate answer.

Devidenko's method is to solve the differential equation (3) corresponding to (4), namely,

$$(1 + 2y)\frac{dy}{d\lambda} = 1 \tag{9}$$

with two possible initial conditions, either $y(0) = 0$ or $y(0) = -1$, which are the solutions of (4) at $\lambda = 0$. The exact solutions y_1 and y_2 in (5), (6) correspond to the integration from 0 and -1, respectively. The integration has no difficulty in integrating (9) from $\lambda = 0$ to $\lambda = 1$, although the perturbation series break down at $\lambda = 1/4$.

Now consider the use of a Newton-Raphson method (without continuation) to solve the equation (2). This method calculates successive approximations $x_k(\lambda)$ from the recursive formula

$$x_{k+1}(\lambda) = x_k(\lambda) - \frac{\partial G}{\partial x}[x_k(\lambda),\lambda]^{-1}\, G[x_k(\lambda),\lambda] \tag{10}$$

To start the method one requires an initial estimate $x_1(\lambda)$. In essence, if $x_1(\lambda)$ is "reasonably close" to the exact solution $x(\lambda)$, then the estimates $x_k(\lambda)$ converge quickly to $x(\lambda)$ as k tends to infinity. The disadvantage of using Newton-Raphson's method is that we do not have an initial $x_1(\lambda)$ which is reasonably close to the exact solution $x(\lambda)$. A continuation method such as Davidenko's can help overcome this difficuly.

As a specific example, consider the trivial equation

$$1 - e^{-(x-\lambda)} = 0 \tag{11}$$

and suppose that we know that $x(10) = 10$ but we wish to approximate $x(0)$ from this initial guess. The guess $x_1(0) = 10$ is disastrous for applying Newton-Raphson's method. The recursion equation (10) for $\lambda = 0$ becomes

$$x_{k+1} = x_k + 1 - e^{x_k} \tag{12}$$

and we find that if $x_1 = 10$, then $x_2 \simeq -22015$, $x_3 \simeq -22014$, $x_4 \simeq -22013$, and so on. One requires over 22,000 iterations of (12) to approach the true root $x = 0$.

Davidenko's method leads to the numerical integration of the differential equation

$$\frac{dx}{d\lambda} = 1$$

with the initial condition $x(10) = 10$. If we use Euler's method of integration with step size h, we will require roughly $10 \times h^{-1}$ integration steps to reach the approximation of $x(0)$. Note that the approximation will be accurate to order h. Thus, Davidenko's method leads to the approximate solution of (11) at $\lambda = 0$ in a reasonably few integration steps if we do not require great accuracy in our approximation.

This trivial example, although extremely exaggerated, does illustrate the basic distinctions between a Newton-Raphson local technique and a Davidenko global technique for solving nonlinear equations. The Newton-Raphson technique is a fast converging approximation method provided that the initial estimate is close to the exact solution. In this case, the error at each step is reduced by squaring. That is, the error δ_{k+1} at the k+1 iteration is approximately δ_k^2, and thus convergence is very fast. A continuation method such as Davidenko's method is inferior to Newton-Raphson's method from the standpoint of accuracy with respect to computation speed. The Davidenko method makes a final error proportional to the integration step size h in the integration of (3) (or proportional to h^{α} for $\alpha > 1$ if a better integration scheme than Euler's method is employed). If one tries to obtain good accuracy by choosing h small, the integration time may be quite long. For example, numerically integrating (13) to obtain the solution $x(0)$ of (11) to four decimal places accuracy would require about 10^5 integration steps using Euler's method of integration.

However, although the continuation methods are inefficient in computing highly accurate solutions, these methods are much less sensitive than Newton-Raphson's to the initial estimate of the solution.

Similarly, continuation methods are not as efficient as perturbation expansions when the perturbation parameter is small. On the other hand, continuation methods such as Davidenko's method provide solutions even when the perturbation parameter is large.

Ideally, one should use a continuation method together with a method such as Newton-Raphson or perturbation analysis. For example, one might first use Davidenko's method with a modest integration step size to compute an approximation reasonably close to the exact solution. Then, using Newton-Raphson's method and using the Davidenko approximation as an initial estimate, one could obtain a very accurate approximation of the exact solution. Current research indicates that by using Davidenko's method as a predictor step and the Newton-Raphson method as a corrector, one can obtain an approximation algorithm which is much more efficient than either and which converges to the desired solution from even poor initial estimates. The reader should refer to Rheinboldt [4], Wacker, Zarzer, and Zulehner [5], and Den Heijer [6] for more detail. Of course, for any particular application the efficiency of the algorithm will depend on intelligent choice of the continuation parameter.

B. Bifurcation and Singular Perturbation Phenomena in Continuation

The method of differentiation with respect to a parameter discovered by Davidenko [3] solves the equation

$$G(x,\lambda) = 0 \tag{14}$$

by integrating the differential equation

$$\frac{\partial G}{\partial x}(x,\lambda)\frac{dx}{d\lambda} + \frac{\partial G}{\partial \lambda}(x,\lambda) = 0 \tag{15}$$

from an initial condition $x(\lambda_0) = x_0$ where

$$G(x_0,\lambda_0) = 0$$

The integration in (15) may proceed as long as the derivative $(\partial G/\partial x)[x(\lambda),\lambda]$, which is a square matrix in the finite dimensional case, remains nonsingular and may be inverted to solve for $dx/d\lambda$ in (15). If $(\partial G/\partial x)[x(\lambda),\lambda]$ should become singular for some value of $\lambda = \lambda_1$, then a sudden change occurs in the number and nature of solutions of (14) for λ near λ_1.

Basically, three different types of behavior can occur. Let $x_1 = x(\lambda_1)$ and suppose $(\partial G/\partial x)(x_1,\lambda_1)$ is singular. We say (x_1,λ_1) is a *limit point* (also called a *weak singularity*) if the $n \times (n + 1)$ rectangular matrix $[(\partial G/\partial x)(x_1,\lambda_1),\ (\partial G/\partial \lambda)(x_1,\lambda_1)]$ has full rank n. In this case the solution of (14) looks like Figure 1 for (x,λ) near (x_1,λ_1). Using methods of Keller [7] or Kubicek [8], it is fairly easy to follow the curve $[x(\lambda),\lambda]$ around the limit point.

The second kind of singularity occurs when $[(\partial G/\partial x)(x_1,\lambda_1),\ (\partial G/\partial \lambda)(x_1,\lambda_1)]$ has rank less than n. In this case we call (x_1,λ_1) a *bifurcation point* (or *strong singularity*). In this case the solution of (14) typically looks like Figure 2 for (x,λ) near (x_1,λ_1). Under certain conditions (see Crandall and Rabinowitz [9]), there are only two curves intersecting at (x_1,λ_1). This situation is called *simple bifurcation*.

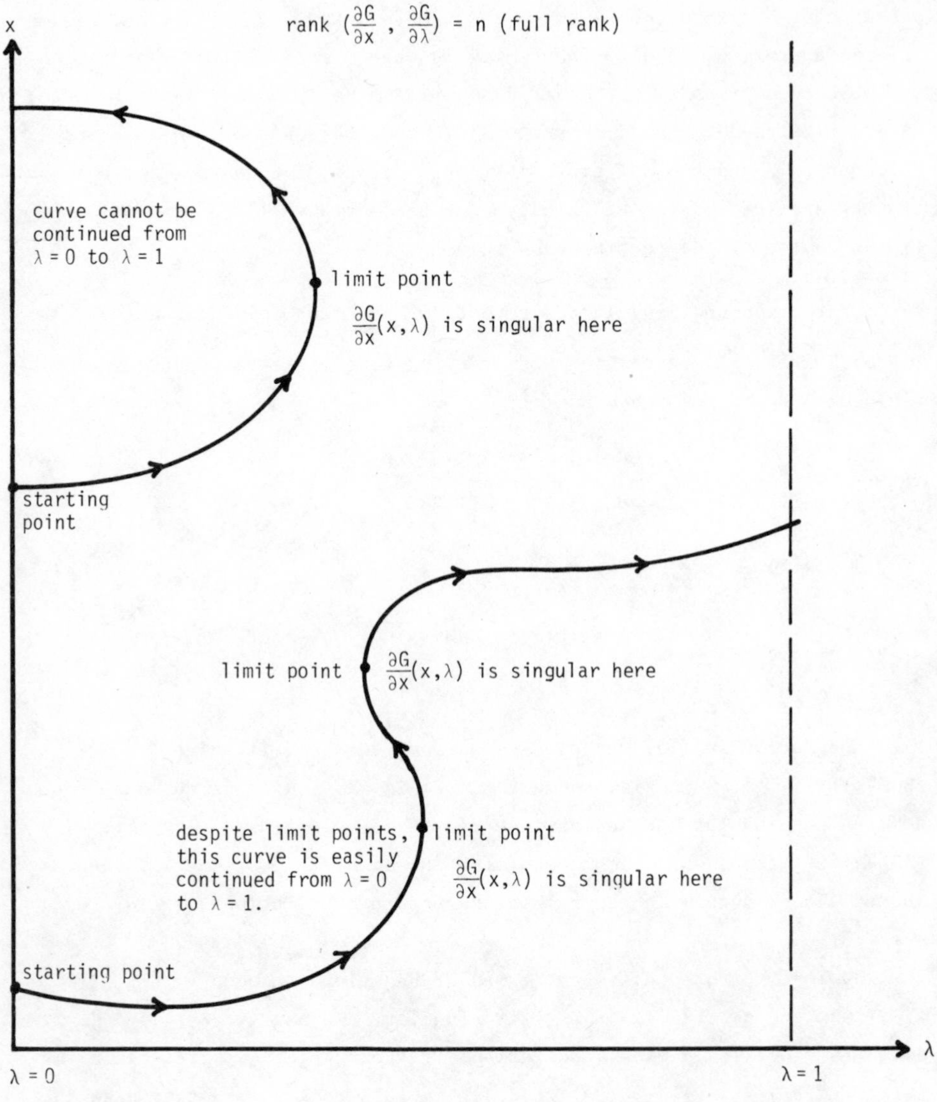

Figure 1 Two Examples of Continuation Curves with Weak Singularities or Limit Points for $G(x,\lambda) = 0$. (Courtesy DOE.)

Note that in case x is an infinite dimensional vector, one may have infinitely many curves intersecting at a bifurcation point. This situation occurs, for example, when singular arc solutions exist for trajector optimization problems. In the finite dimensional case several algorithms exist for dealing with simple bifurcations. See Rheinboldt [10], for

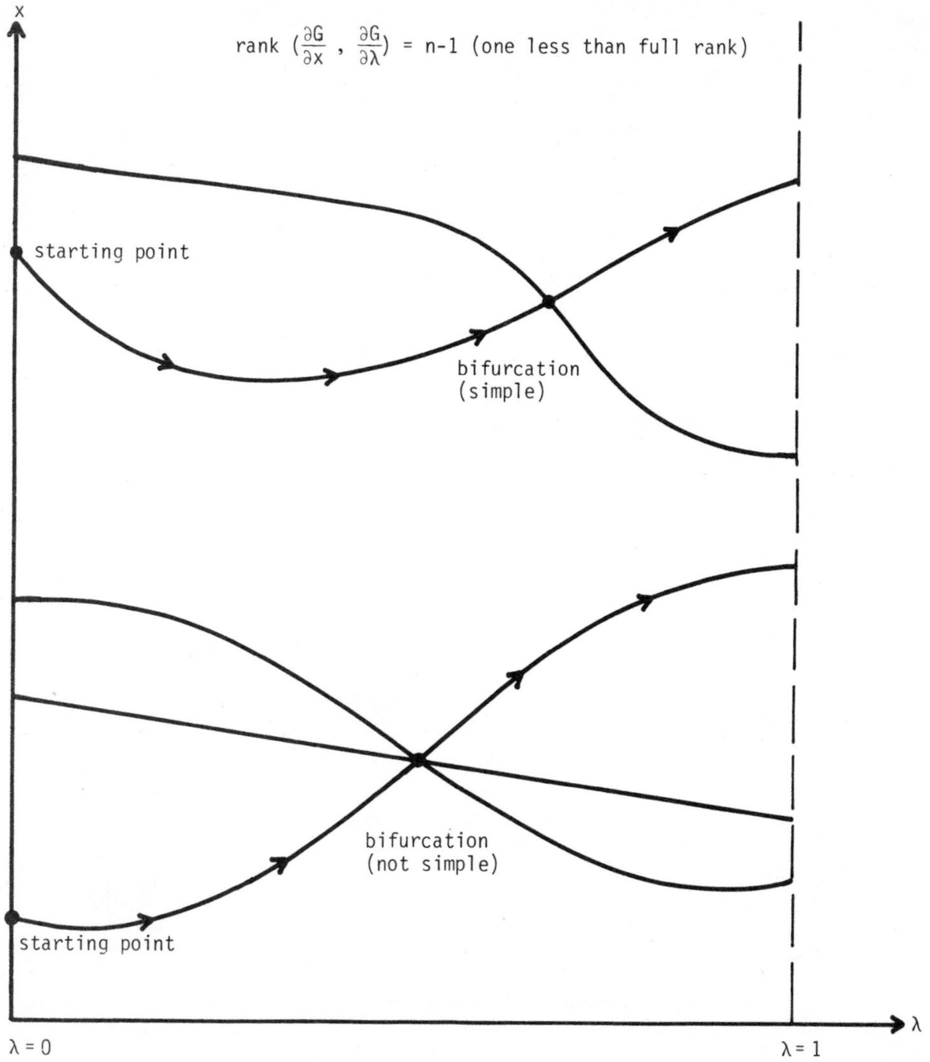

Figure 2 Examples of Simple and Nonsimple Bifurcation of $G(x,\lambda) = 0$. (Courtesy of DOE.)

example, if the bifurcation is not "simple" but the rank of $[(\partial G/\partial x), (\partial G/\partial \lambda)]$ is $n - 1$, one can also use a numerical approximation of the Lyapunov-Schmidt transformation (see Crandall and Rabinowitz [9]) to reduce the problem (14) near (x_1,λ_1) to solving one scalar equation in two unknowns. If x has small dimension one can also use direct search techniques near the bifurcation (x_1,λ_1) to solve (14).

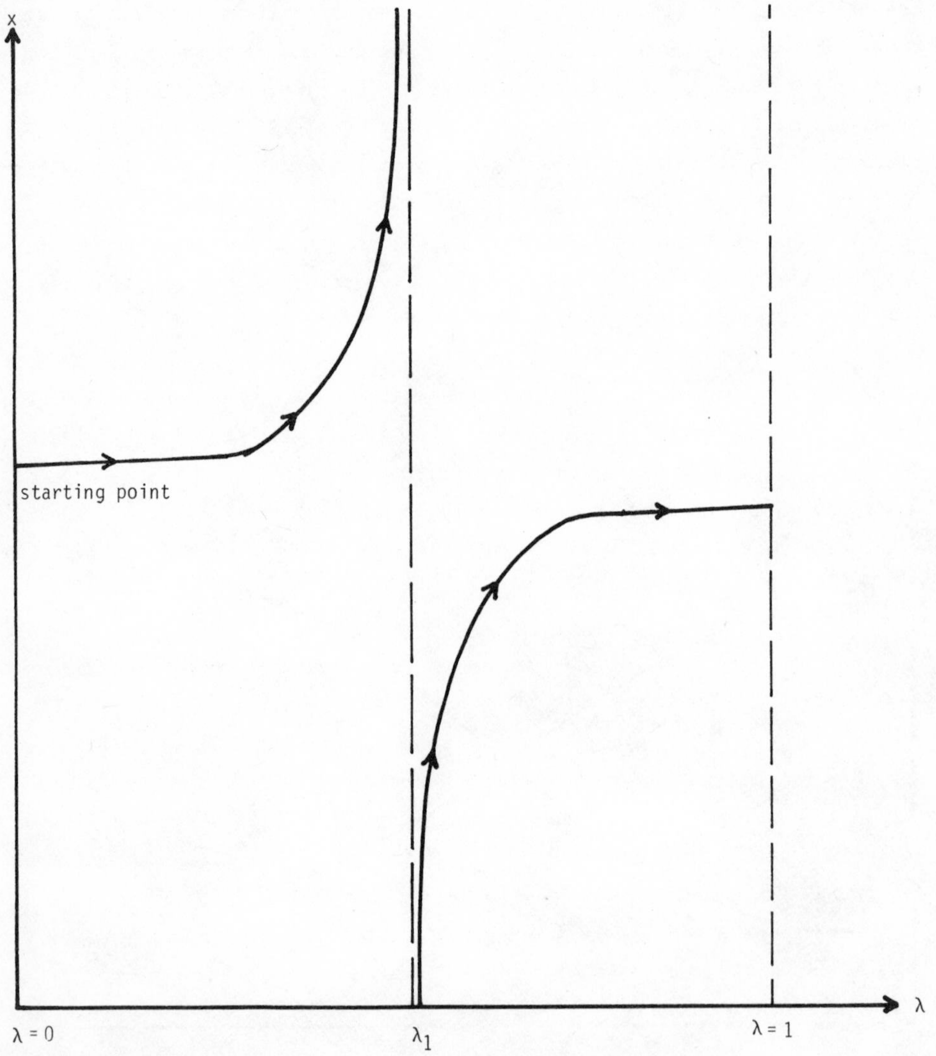

Figure 3 Singular Perturbation Bifurcation of $G(x,\lambda) = 0$. (Courtesy DOE.)

The third type of singularity is a singular perturbation bifurcation which occurs when the solution curves $x(\lambda)$ of (14) become infinite as $\lambda \to \lambda_1$ (see Figure 3). This type of singularity has been studied much less than the previous two; however, see Matkowsky and Reiss [11] for some results.

The proper treatment of singularities is extremely important in applying the continuation method to the solution of nonlinear problems. Either one must treat the singularity so that the solution curve $x(\lambda)$ can be continued through it, as we mentioned above, or one must choose the function G in (14) so that no singularities occur. Some recent work, using methods of differential topology, shows that in many problems there is a large class of G for which no bifurcations or singular perturbation bifurcations can occur. See the recent papers of Alexander [12] and Watson [13] for further information and references. Note that one attractive feature of continuation methods is the possibility of obtaining all of the multiple solutions of a nonlinear problem by continuing the solution curves $x(\lambda)$ from several initial points. In this case, it may be useful to choose G intentionally so that it has bifurcation points. Figure 4 is meant to be optimistiaclly suggestive. Note that in the case where $G = 0$ represents a system of polynomial equations, Garcia and Zangwill [14] have presented a continuation method which determines all of the multiple solutions. See also Garcia and Zangwill [15] for an extension to more general cases.

C. Brief History and Current Status of Continuation Methods

Continuation methods (also called imbedding methods and homotopy methods) have a long history in mathematics and numerical analysis, going back to the last century at least. The basic idea of the method is very simple and it seems to be continuously rediscovered by independent researchers. A good discussion of continuation methods from the theoretical numerical analysis point of view occurs in Ortega and Rheinboldt [16]. Another good reference with emphasis on applications is Wasserstrom [17].

The use of a continuation method as a numerical method begins with Lahaye (1934) [2], who used Newton's method in a continuation method to solve nonlinear equations. He did most of his work for a single equation in a single unknown. In the 1950s Davidenko [3] used differentiation with respect to a parameter to solve finite systems of nonlinear equations and applied the technique to several numerical problems. In the same vein, but more recently, Rheinboldt (1975 - 1979) [4,10] has worked to develop an adaptive continuation program incorporating both Lahaye's and Davidenko's methods. Thus, to make the continuation method efficient, Rheinboldt is developing algorithms to choose the step size for the predictor and the precision for the corrector adaptively. The step size and precision become small only near singular points and near the desired value of the continuation parameter, where more accuracy is required than elsewhere.

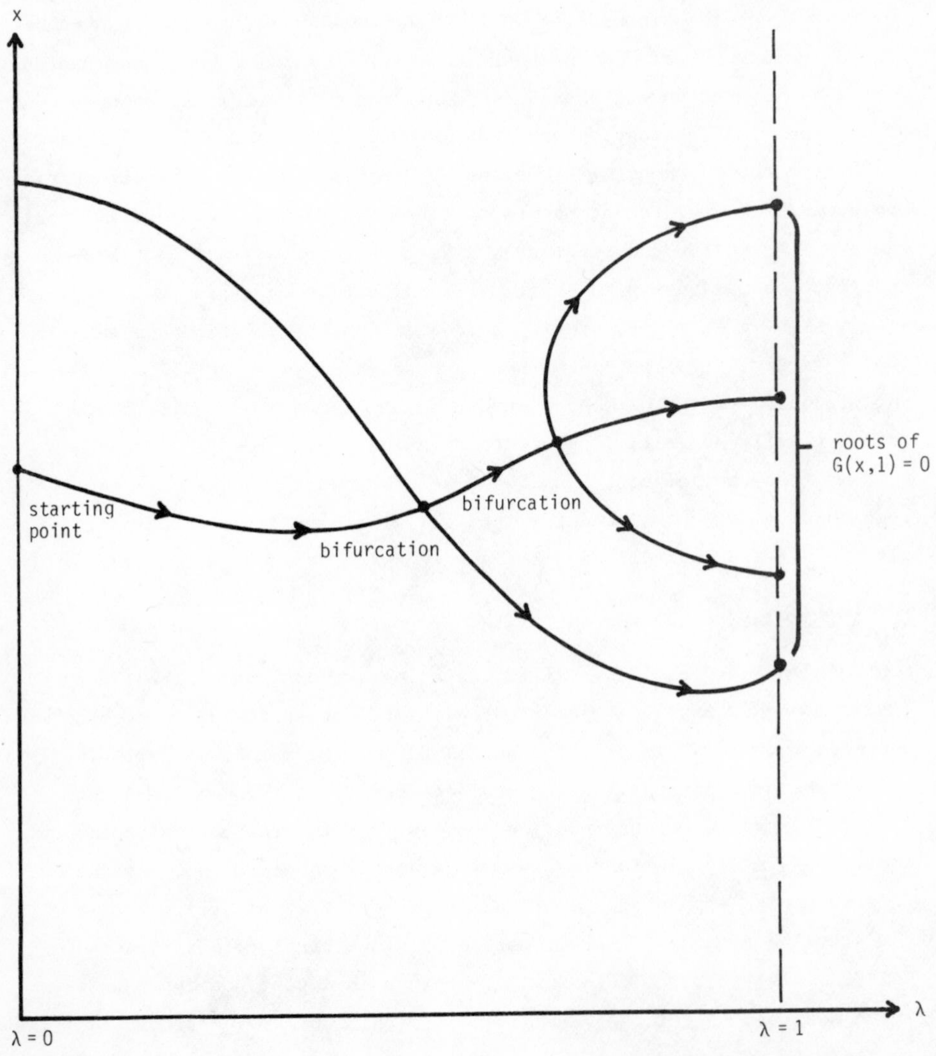

Figure 4 Using Bifurcations to Obtain Multiple Solutions of $G(x,\lambda) = 0$ for $\lambda = 1$. (Courtesy DOE.)

See also the recent collection of papers by Wacker [18] and the book of Den Heijer [6] for similar work.

A different approach to continuation has recently developed in the work of Yorke, Li, Kellogg, Mallet-Paret, Watson, and others (see Watson [13]). This work originates in the abstract field of differential topology and the technique is called a *homotopy method* (the function G in (14) is called a homotopy in topology). The basic result of this approach shows that in certain circumstances if one chooses G in (14) at "random" from a simple family of possible G's, then "with probability one" the chosen G will have no bifurcations or singular perturbation bifurcations. At worst it will have limit points. Moreover, one is guaranteed that the differential equation (15) can be integrated to find G over all λ parameter values. In other words, "almost all" G work in some cases. At present, work is being done to determine what these theoretical results mean in practice.

Application of continuation methods to solving trajectory optimization methods is carried out by using a continuation method to solve the two-point boundary value problem (TPBVP) corresponding to the Euler-Lagrange necessary conditions for optimality. Roberts and Shipman [19] and Kubicek and Hlavacek [20] use continuation methods to solve TPBVP's. The papers of Montgomery [21] and Jamshidi [22] specifically treat control problems by continuation of a TPBVP. Note that we are not limited to using continuation on the TPBVP but can use it directly on the optimization problem together with some local optimization technique like Fletcher-Powell, conjugate gradient, etc.

The references we have given here are far from complete and they do not necessarily represent the major works on continuation methods. See Mehra, et al. [23] for an extensive bibliography of continuation method papers. Also see the survey article by Wacker [18] and the recent book by Den Heijer [6] for some additional references.

III. APPLICATION TO STABILITY AND OPTIMIZATION PROBLEMS

A. Introduction

In this section we describe applications of continuation methods to the bifurcation and stability analysis of supersonic aircraft and electric power system models, to the optimization of aircraft trajectories and to the computation of limit cycles in the study of aircraft spin. The continuation algorithm that was employed in these applications is a combination and variation of the algorithms and results of Kubicek [8], Keller [7], and Crandall and Rabinowitz [9]. The algorithm is of the predictor-corrector type, although it is not adaptive in the sense of Wacker, Zarzer, and Zulehner [5]. It easily handles weak singularities, and can also handle strong singularities with somewhat less efficiency. General bifurcations are treated by means of an approximate Lyapunov-Schmidt transformation (see [9]) coupled with a direct search technique in a reduced dimensional space. In the case of simple bifurcations we can use an approximate version of the Crandall and Rabinowitz [9] scheme, provided that second derivative information is available.

B. Bifurcation and Stability Analysis

The abstract problem of bifurcation and stability analysis of dynamic systems is to analyze how the trajectories of

$$\dot{x}(t) = f[x(t),u(t)] \tag{16}$$

change as the parameter u(t) varies. If u(t) varies slowly, so that $\dot{u}(t)$ is small in magnitude, then except for certain special values of x and u we can approximate (16) by assuming that the input u(t) is constant in time. The special values of x and u are those for which

$$0 = f(x,u)$$

and $(\partial f/\partial x)(x,u)$ has some imaginary eigenvalues with all other eigenvalues having negative real parts. In other words, these values of x and u determine the onset of instability in the dynamic system defined by (16). The set of such x and u is called the bifurcation surface and knowledge of its location is valuable for understanding the dynamics and especially the stability behavior of (16).

To see how continuation methods help solve this problem, consider the special case where f(x,y) is given by a potential function so that

$$f(x,u) = \frac{\partial \phi}{\partial x} (x,u) \tag{17}$$

From (17) it follows that

$$\frac{\partial f}{\partial x} (x,u) = \frac{\partial^2 \phi}{\partial x^2} (x,u) \tag{18}$$

where $\partial^2\phi/\partial x^2$ is the matrix of second order partial derivatives of the scalar function ϕ. Since this matrix is symmetric and real-valued, all eigenvalues of $\partial f/\partial x$ must be real. Consequently, the only imaginary eigenvalues are ones equal to 0 and the conditions for the bifurcation points (x,u) are

$$0 = f(x,u) \tag{19}$$

$$0 = \det \left[\frac{\partial f}{\partial x} (x,u)\right] \tag{20}$$

Equations (19) and (20) are used in a continuation algorithm to find the solutions (x,u). In practice, all but two components of the control $u = (u_1, u_2, \ldots, u_m)$ are fixed, and one of the remaining two control components, say u_1, is varied. The other control component, say u_2, and the components of x are then determined from (19),(20). In this way, we construct a curve in (x, u_1, u_2) space. By repeating the procedure for several fixed values of $(u_3, u_4, \ldots, u_m)$, we can build up an approximate picture of the bifurcation surface in (x,u) space. Fortunately for the sake of computational efficiency, many interesting problems can be reduced to considering just two or three component controls u.

One example of the application of this bifurcation and stability analysis occurs in the study of electric power system steady-state stability and related load-flow equations. A simple example is given by the three machine system shown in Figure 5 where T_1, T_2, and T_3 represent the equivalent electric torque applied at terminals 1, 2, and 3. The torques are related to the power via the relation $T_i = P_i/\omega$, where ω is the synchronous frequency; $F_{ij} = F_{ji} = V_i V_j/\omega^2 L_{ji}$ denotes the maximum torque that can be transferred between machines i and j, where V_i denotes the voltage at i and L_{ij} is the reactance between i and j. Let δ_i denote the phase angle of V_i and define internode angles (α_1, α_2) with respect to machine 3 as

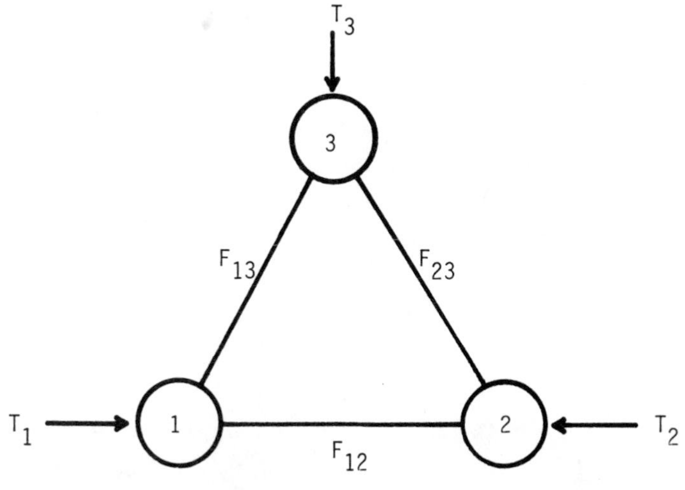

Figure 5 Three Machine Example. (Courtesy DOE.)

$$\alpha_1 = \delta_1 - \delta_3 \qquad (21)$$
$$\alpha_2 = \delta_1 - \delta_2$$

The equilibrium equations of the system, neglecting damping and transfer conductances, can be written as

$$F_1 = F_{12} \sin\alpha_1 + F_{12} \sin(\alpha_1 - \alpha_2) \qquad (22)$$
$$T_2 = F_{23} \sin\alpha_2 + F_{12} \sin(\alpha_2 - \alpha_1)$$

Also, $T_1 + T_2 + T_3 = 0$. There are two state variables (α_1,α_2) and five parameters or control variables $(T_1,T_2,F_{13},F_{23},F_{12})$ in this problem. We will examine the effect of varying these parameters on the solutions of (22). This is important for transient stability analysis since the disturbance experienced by the system would result in a sudden change in these parameters. Furthermore, the transient stability limit and the mode of transient instability would also depend on the values of these parameters. The analysis of (22) in the augmented space of state and control variables is also essential for a study of the structurally stable bifurcations.

The effect of varying only T_1, keeping T_2 fixed at zero, is shown in Figures 6 and 7. Notice that (22) has 6, 4, 2, or 0 solutions in the range $-\pi < \alpha_1,\alpha_2 \leq \pi$ depending on the value of T_1. There is, however, only one stable solution in this simple case and this is denoted by S. The nature of the singular point is determined by the eigenvalues of the Jacobian of (22). There are six bifurcation points (B.P.) separating the intervals on the T_1-axis with different numbers of equilibrium solutions. The significance of the B.P.'s is that the Jacobian is singular at these points and that the stability characteristics of the equilibrium point change at B.P. Figures 6 and 7 also show the value of the potential energy difference, V, between the saddle or unstable equilibrium points and the corresponding stable equilibrium points. Saddles are also unstable equilibrium points, but the term "unstable" is used here to denote points with positive eigenvalues corresponding to maxima of the potential functions.

A second example can be found in the analysis of aircraft stability at high angles of attack. The angle of attack α is shown in Figure 8 and is the angle between the body axis X_b and the stability axis X_s. The report of Mehra, Kessel, and Carroll [24] contains more details for this

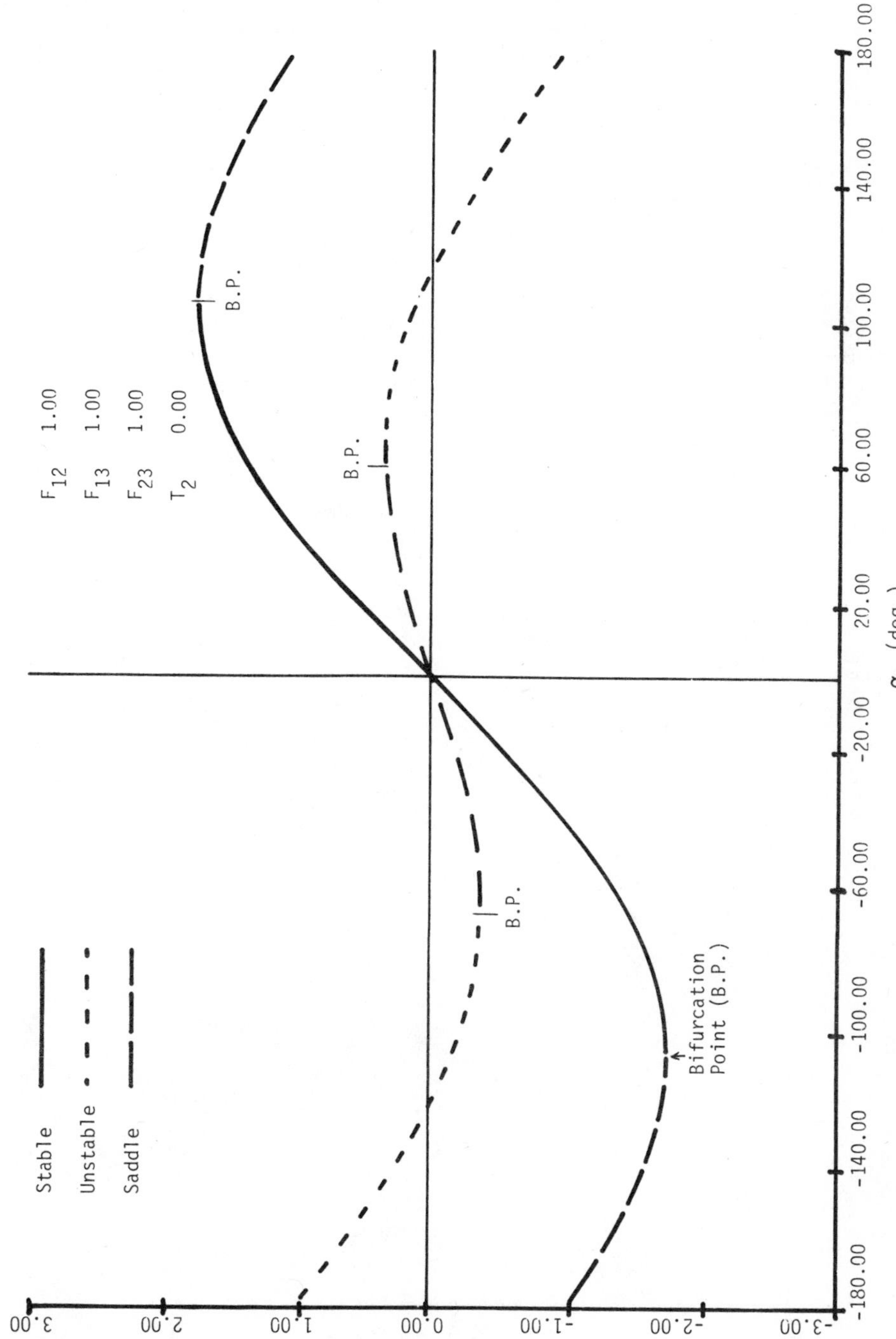

Figure 6 Torque Angle α vs. Electric Torque T_1 for a Three-machine System. (Courtesy DOE.)

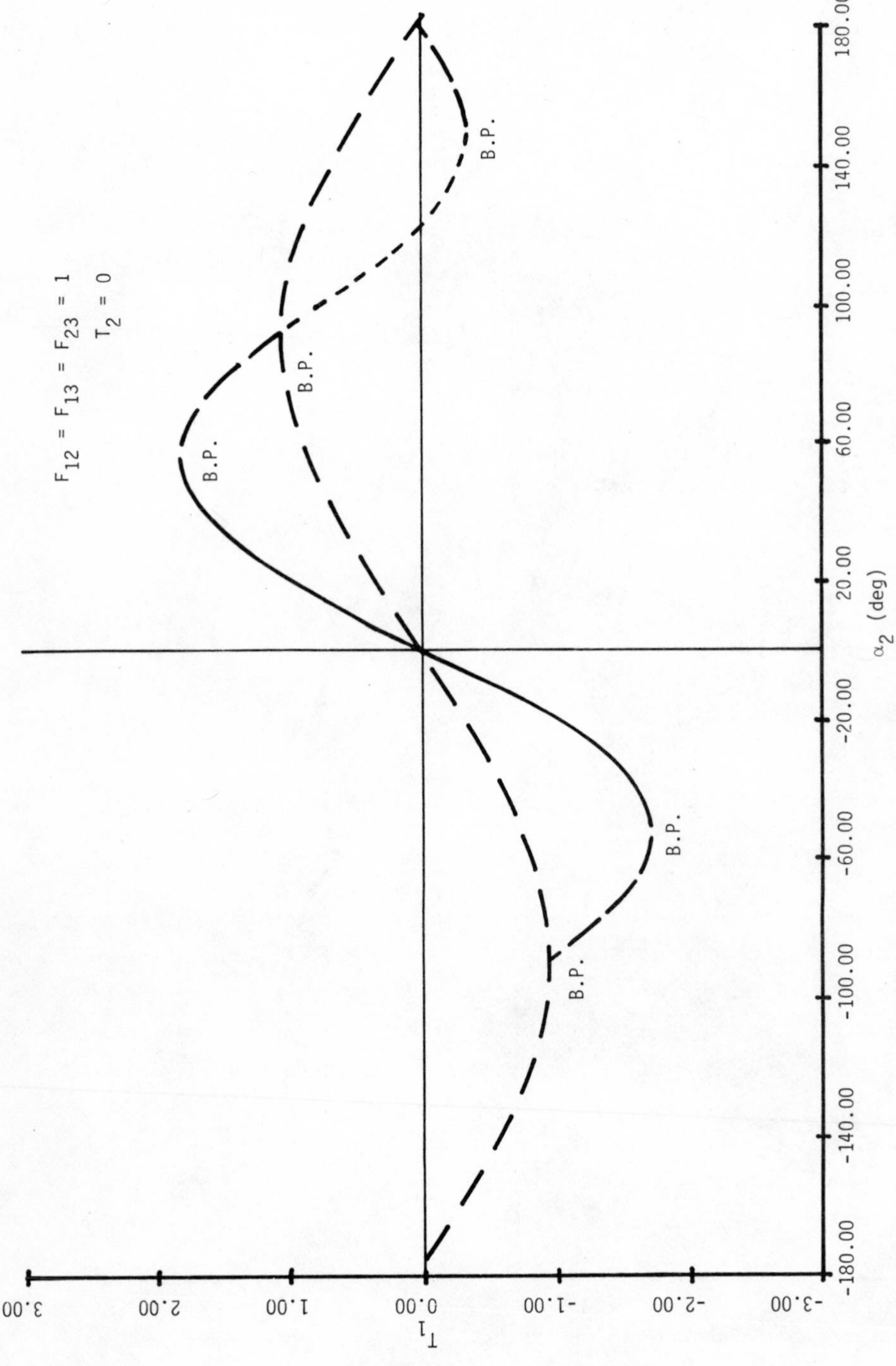

Figure 7 Torque Angle α_2 vs. Electric Torque T_1 for a Three-machine System. (Courtesy DOE.)

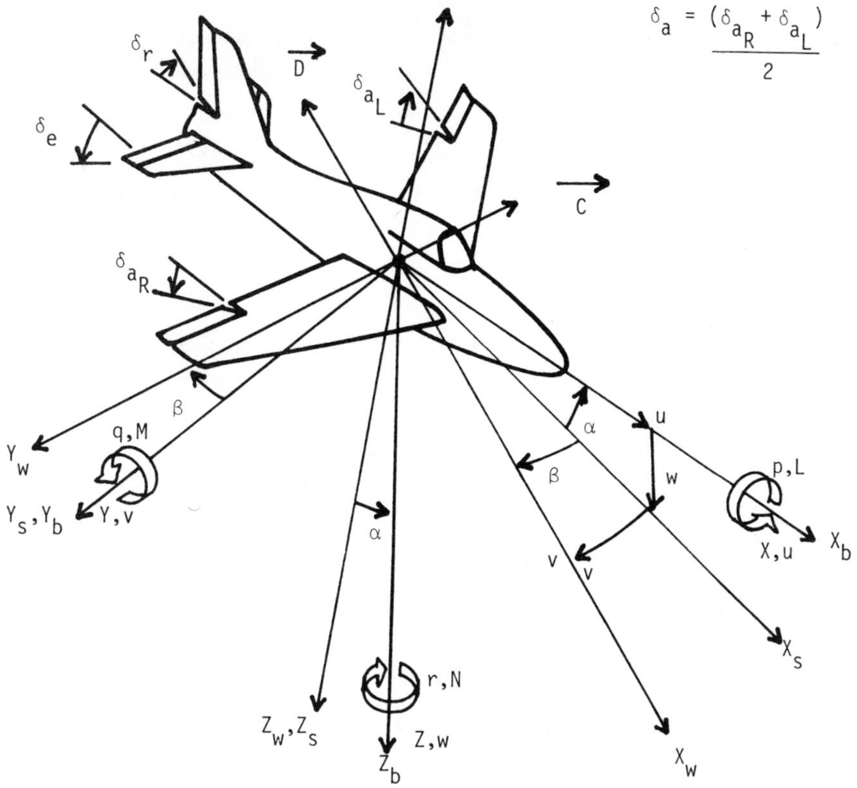

Figure 8 Body System of Axes. Arrows indicate positive direction of quantities. (Reprinted from Office of Naval Research, report ONR-CR215-248-1, June 30, 1977.)

problem. High angle-of-attack phenomena have been of interest to aerodynamicists, aircraft designers, pilots, and control system analysts ever since the advent of modern high performance aircraft. Due to the concentration of inertia along the fuselage, the modern jet fighters are highly susceptible to post-stall departures and spin. Extensive wind tunnel testing and radio-controlled flight testing has been done over the last 20 years to gain better understanding of the dynamic instabilities at high angles of attack. A basic problem has existed in interpreting these data and in making predictions of aircraft dynamic behavior so as to achieve close agreement with flight test data.

Aircraft dynamic behavior at high angles of attack is highly nonlinear, and in the past there has been a lack of suitable techniques for analyzing the global behavior of nonlinear systems. Under an ongoing project we have developed a new approach based on bifurcation analysis and using continuation algorithms for computation. The approach has been applied to specific jump, hysteresis, and limit cycle phenomena such as roll-coupling, pitch-up, wing rock, buffeting, departure, and divergence. Three different aircraft have been considered for comparison purposes, and it has been shown how different types of instabilities and families of limit cycles arise as the control variables are varied. A complete representation of the aircraft equilibrium and bifurcation surfaces is given in an eight-dimensional space consisting of roll rate, pitch rate, yaw rate, angle of attack, sideslip angle, elevator, aileron, and rudder deflections. Two-dimensional projections of the equilibrium and bifurcation surfaces provide pictorial representations of the aircraft global stability and control behavior at high angles of attack.

Mathematically, the problem fits into the abstract framework discussed at the beginning of this section. One simple version of the problem has a five-dimensional state vector $x = (p,q,r,\alpha,\beta)$ and a three-dimensional control vector $u = (\delta e,\delta a,\delta r)$. Here p, q, r are body-axis coordinates of angular momentum, α is angle of attack, and β is sideslip angle. The controls δe, δa, δr represent angular deflections for elevator, aileron, and rudder, respectively. The differential equations in this case are

$$\dot{p} = \ell_\beta \beta + \ell_{\alpha\delta a}\Delta\alpha\delta a + \ell_q q + \ell_r r + \ell_{\beta\alpha}\beta\Delta\alpha + \ell_p p - i_1 qr + \ell_{\delta a}\delta a + \ell_{\delta r}\delta r \tag{23}$$

$$\dot{q} = \bar{m}_\alpha \Delta\alpha + \bar{m}_q q + i_2 pr + m_{\delta e}\delta e - m_{\dot{\alpha}} p\beta \tag{24}$$

$$\dot{r} = n_\beta \beta + n_{\alpha\delta a}\Delta\alpha\delta a + n_r r + n_p p + n_{p\alpha} p\Delta\alpha - i_3 pq + n_{\delta a}\delta a + n_{\delta r}\delta r \tag{25}$$

$$\dot{\alpha} = q - p\beta + z_\alpha \Delta\alpha + z_{\delta e}\delta e \tag{26}$$

$$\dot{\beta} = y_\beta \beta + p(\sin\alpha_0 + \Delta\alpha) - r\cos\alpha_0 + y_{\delta a}\delta a + y_{\delta r}\delta r \tag{27}$$

The physical definition of the terms in this equation and its derivation can be found in Mehra, et al. [24]. However, let us note here that many terms such as ℓ_β, $\ell_{\alpha\delta a}$, ℓ_q, etc. have nonlinear dependence on the state and control variables which is not explicit in equations (23-27). One difference between this aircraft stability problem and the previously discussed power system problem is that equations (23-27) do not come from a potential function. Thus, there are in general complex eigenvalues instead of just real eigenvalues. One interesting case occurs when a conjugate pair of complex eigenvalues crosses the imaginary axis as the control parameter u varies. This corresponds to a Hopf bifurcation, as discussed in Marsden and McCracken [25], and indicates the onset of physically important limit cycle behavior (i.e., spin).

C. Two-point Boundary Value Problems: Trajectory Optimization and Limit Cycles

Continuation methods can be used to solve two-point boundary value problems (TPBVP) of the following type:

$$\frac{\partial x}{\partial t}(t,\lambda) = f[x(t,\lambda),t,\lambda] \tag{28}$$

$$x(0,\lambda) = z(\lambda) \tag{29}$$

$$\phi\{a(\lambda),\ x[\tau(\lambda),\lambda],\ \tau(\lambda),\ \lambda\} = 0 \tag{30}$$

where $0 \leq t \leq \tau(\lambda)$ and λ is the continuation parameter. If $x(t;a,\lambda)$ denotes the solution of

$$\frac{\partial x}{\partial t}(t;a,\lambda) = f[x(t;a,\lambda),t,\lambda] \tag{31}$$

$$x(0;a,\lambda) = a \tag{32}$$

then the continuation problem is to solve

$$G(a,\lambda) = 0 \tag{33}$$

where

$$G(a,\lambda) = \phi\{a,\ x[\tau(\lambda);a,\lambda],\ \tau(\lambda),\ \lambda\} \tag{34}$$

Many problems fit into this framework, including problems of computing limit cycles and finding optimal trajectories. For example, limit cycles are determined by the terminal condition

$$x[\tau(\lambda),\lambda] - a(\lambda) = 0 \tag{35}$$

so that $\phi(a,x,\tau,\lambda) = x - a$ in (30). We have used this procedure to determine limit cycle trajectories in the study of aircraft at high angles of attack mentioned in Section II.B.

We can also apply the method of continuation to solve the TPBVP which results from the first order necessary conditions for a trajectory optimization problem. The following example is a trajectory optimization problem for a simplified missile interceptor taken from the paper of Schneider and Reddy [26]. Although the aerodynamic model is extremely simplified, this problem is a good example with which to test the continuation algorithm. The state equations in this problem are nonlinear and fourth order, resulting in a nonlinear, eighth order TPBVP. The nonlinear, fourth order state equations are

$$\frac{dx_1}{dt} = x_3 \tag{36}$$

$$\frac{dx_2}{dt} = x_4 \tag{37}$$

$$\frac{dx_3}{dt} = -\lambda x_3 v e^{-x_2/h_s} + u_1 \tag{38}$$

$$\frac{dx_4}{dt} = -\lambda x_3 v e^{-x_2/h_s} + u_2 - 9.81 \tag{39}$$

where $v = \sqrt{(x_3^2 + x_4^2)}$.

In the equations (36-39), x_1 represents a horizontal range variable and x_2 is the height variable. The variables x_3 and x_4 are the velocities corresponding to x_1 and x_2, respectively. The variables u_1 and u_2 are the controls (thrusts) for the problem. The parameter h_s is a height scale which is taken as 6705.6m in this problem. The parameter λ is equal to $(b/2)\rho_s$, where ρ_s is the sea level air density and b is the inverse ballistic coefficient for the missile interceptor modeled by (36-39). The optimality criterion for this problem is the simplest quadratic criterion,

$$\int_0^T \left(u_1^2 + u_2^2\right) dt \tag{40}$$

where T is the fixed final time. The Euler-Lagrange necessary conditions give the following nonlinear, eighth order TPBVP:

$$\frac{dx_1}{dt} = x_3 \tag{41}$$

$$\frac{dx_2}{dt} = x_4 \tag{42}$$

$$\frac{dx_3}{dt} = -\lambda x_3 v e^{-x_2/h_s} - \frac{x_7}{2} \tag{43}$$

$$\frac{dx_4}{dt} = -\lambda x_4 v e^{-x_2/h_s} - \frac{x_8}{2} - 9.81 \tag{44}$$

$$\frac{dx_5}{dt} = 0 \tag{45}$$

$$\frac{dx_6}{dt} = -\frac{\lambda v}{h_s} d^{-x_2/h_s}[x_7 x_3 + x_8 x_4] \tag{46}$$

$$\frac{dx_7}{dt} = -x_5 + \lambda x_7 \left[1 + \left(\frac{x_3}{v}\right)^2\right] ve^{-x_2/h_s} + \lambda x_8 \left[\frac{x_4 x_3}{v^2}\right] ve^{-x_2/h_s} \tag{47}$$

$$\frac{dx_8}{dt} = -x_6 + \lambda x_7 \left[\frac{x_4 x_3}{v^2}\right] ve^{-x_2/h_s} + \lambda x_8 \left[1 + \left(\frac{x_4}{v}\right)^2\right] ve^{-x_2/h_s} \tag{48}$$

where $x_1(0)$, $x_2(0)$, $x_3(0)$, $x_4(0)$, $x_1(T)$, $x_2(T)$ are given and $x_7(T) = x_8(T) = 0$. As in (36-39), $v = \sqrt{(x_3^2 + x_4^2)}$. In equations (41-48), the variables x_5, x_6, x_7, and x_8 are the adjoint variables corresponding to x_1, x_2, x_3, and x_4, respectively. We solved (41-48) by shooting for the missing adjoint initial conditions $x_5(0)$, $x_6(0)$, $x_7(0)$, and $x_8(0)$. The parameter λ (proportional to the inverse ballistic coefficient) was used for continuation from $\lambda = 0$. Note that the $\lambda = 0$ solution corresponds to an interception problem with no air density--i.e., a vacuum solution. Figures 9, 10, 11, and 12 show the range-height trajectories for four values of the ballistic coefficient ranging from ∞ to $3.1266 \times 10^4 \mathrm{Nm}^{-2}$. This corresponds to continuing the parameter λ from 0 (the vacuum solution) to $1.9220 \times 10^{-4} \mathrm{m}^{-1}$. The final time T is 20 sec. and the given initial and final conditions are $x_2(0) = x_3(0) = x_4(0) = 0$, $x_1(0) = -1.524 \times 10^4$m, $x_1(T) = 0$, $x_2(T) = 4.409 \times 10^3$m. These initial and final conditions correspond to the conditions for cases 5 and 6 in Schneider and Reddy [26]. Schneider and Reddy presented the trajectory of case 6 which corresponds to our Figure 12, although the ballistic coefficient for case 6 is about two times that of Figure 12. Figure 11 is very close to case 5, although the ballistic coefficient for Figure 11 is $4.6923 \times 10^4 \mathrm{Nm}^{-2}$, which is slightly less than the coefficient for case 5, namely $5.788 \times 10^4 \mathrm{Nm}^{-2}$. The cost associated with Figure 11 ($\Lambda = 3.524 \times 10^8$) is correspondingly slightly larger than the cost in case 5 ($\Lambda = 3.415 \times 10^8$).

As the parameter λ and the final time T increase, the equations (41-48) become more sensitive to slight changes in the initial values of the adjoint variables. This is partly due to the increased nonlinearity of the problem, but the main trouble comes from the forward integration of the adjoint equations. The adjoint equations are unstable in the forward direction and the differential equations (41-48) will become infinite in a finite amount of time. One solution to the problem which still maintains the shooting method is to use double precision instead of single precision accuracy in computations. Alternatively, one can use one of the function space methods such as quasilinearization as presented in Roberts and

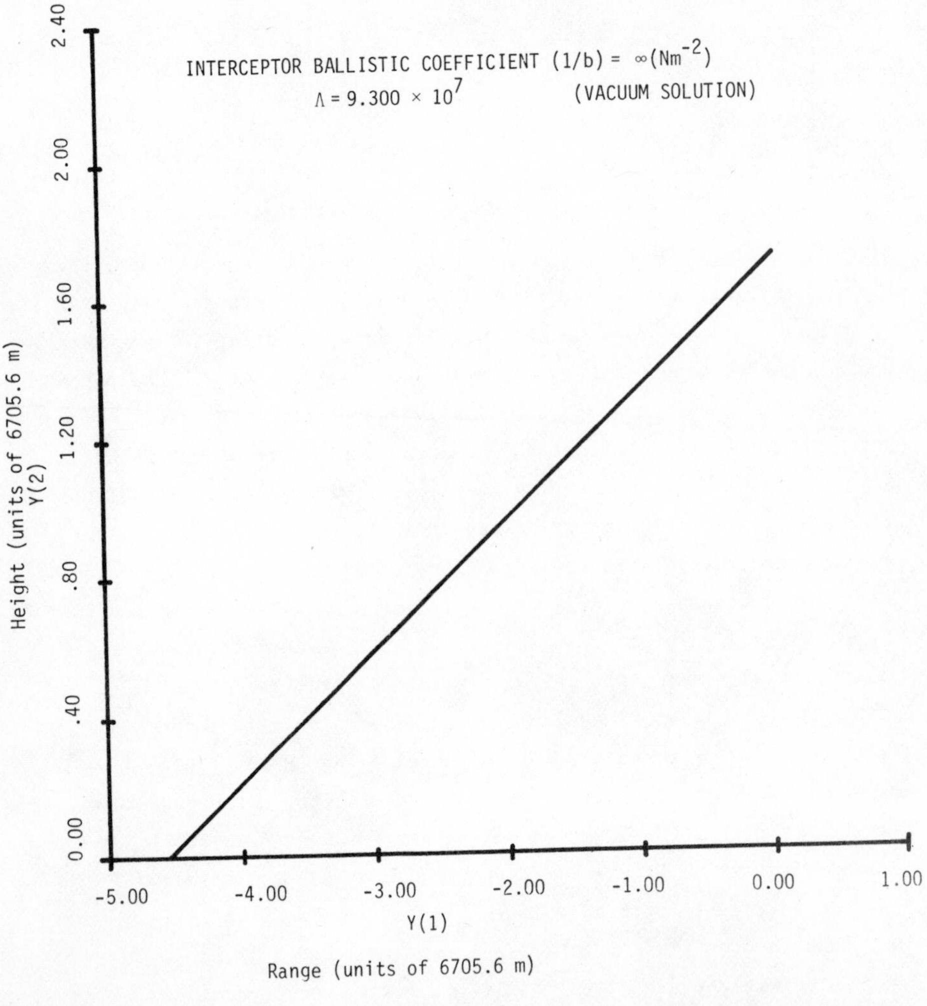

Figure 9 Schneider-Reddy Example. (Reprinted from NASA, Contractor Report 3167, August, 1979.)

Shipman [19] or the back-and-forth shooting method of Orava and Lautala [27,28].

D. Asymptotic Continuation for Trajectory Optimization

Asymptotic methods such as multiple time scale analysis, singular perturbation theory, matched asymptotic expansions, etc., have long been used in applied mathematics and in engineering problems to obtain useful approximations of solutions to trajectory problems. The reader may find many

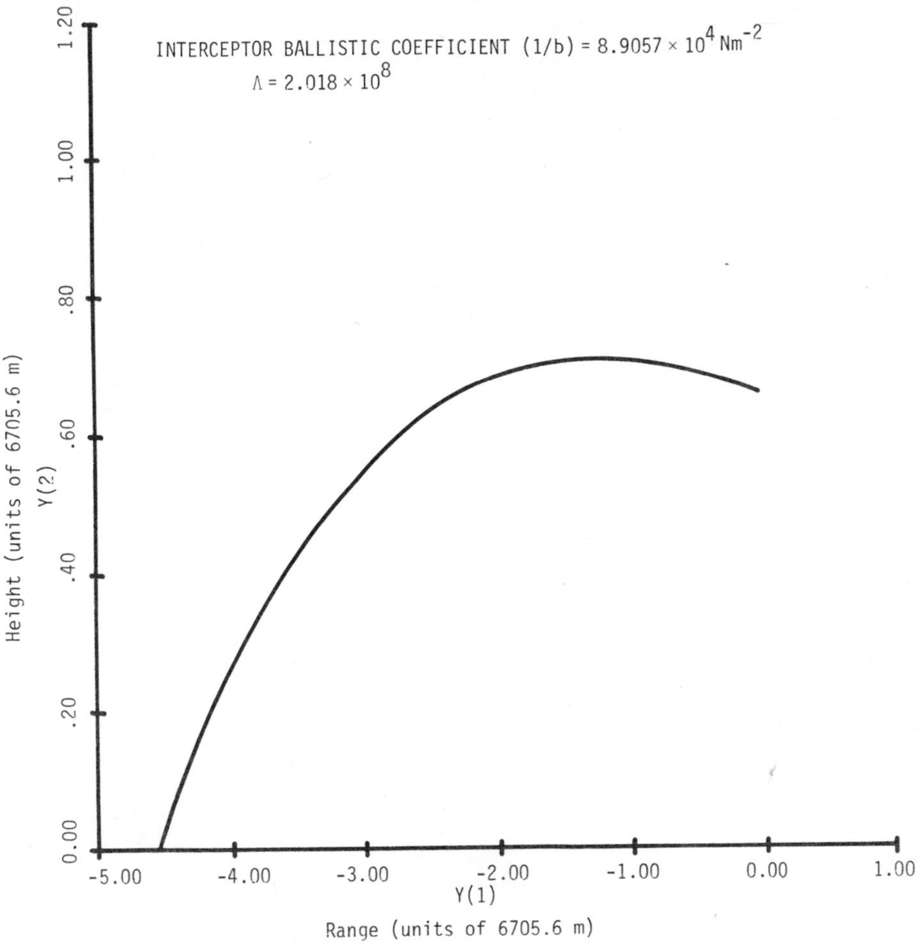

Figure 10 Schneider-Reddy Example. (Reprinted from NASA Contractor Report 3167, August, 1979.)

techniques and examples presented in Nayfeh [29]. An overview of singular perturbation theory applied to the trajectory optimization problems of control theory is given in Kokotovic, O'Malley, and Sannuti [30]. Compare also the application of matched asymptotic expansions in Ardema [31] and the multiple time scale analysis in Ramnath and Sinha [32].

Asymptotic methods are attractive from several points of view. In engineering applications, for example, these methods provide a systematic framework for order reduction and permit one to simplify a higher dimensional model for purposes of analysis and on-line implementation. In addition, when the interaction of fast and slow variables in a problem results

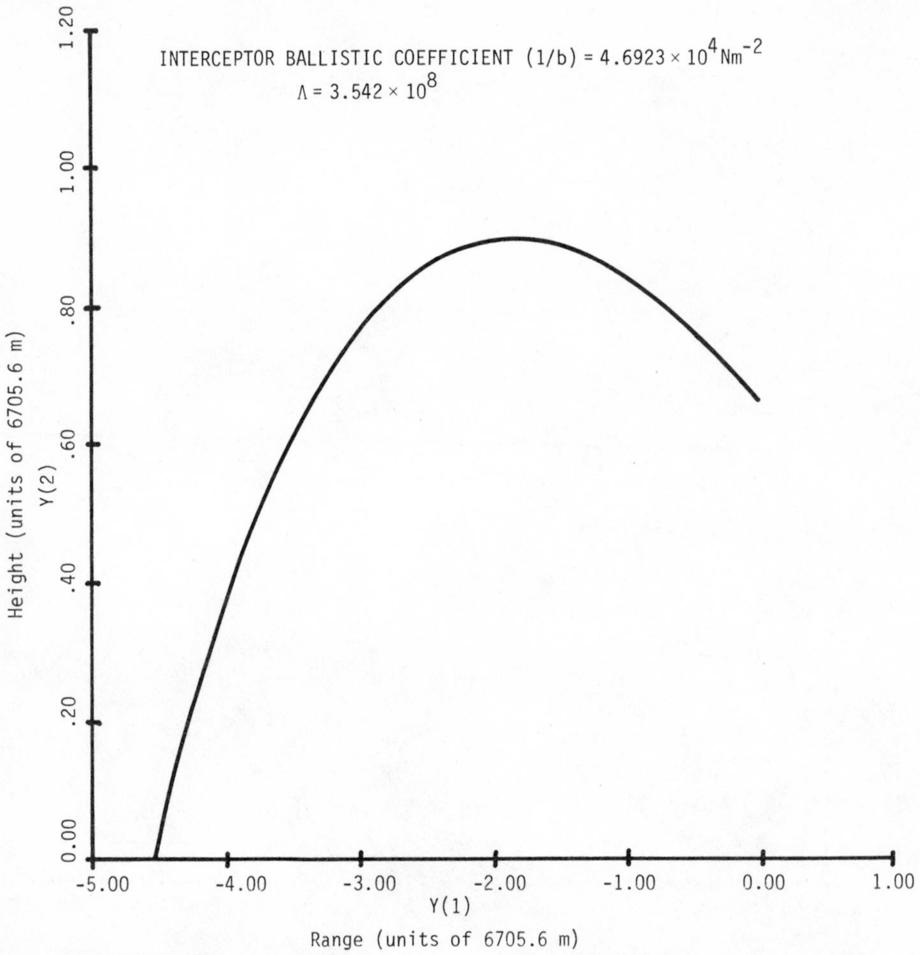

Figure 11 Schneider-Reddy Example. (Reprinted from NASA Contractor Report 3167, August, 1979.)

in a stiff numerical problem, asymptotic approximation methods provide robust numerical approximations to make the computation efficient.

Unfortunately, current methods of asymptotic approximation have certain limitations that seriously impede their direct application to many complex trajectory problems encountered in modern engineering. For example, it sometimes happens that the lowest order asymptotic approximation is not sufficiently accurate and higher order correction terms are needed. These higher order corrections are generally difficult to compute--in anything but a low dimensional problem, one would not usually go beyond the first order correction. Computing correction terms involves taking

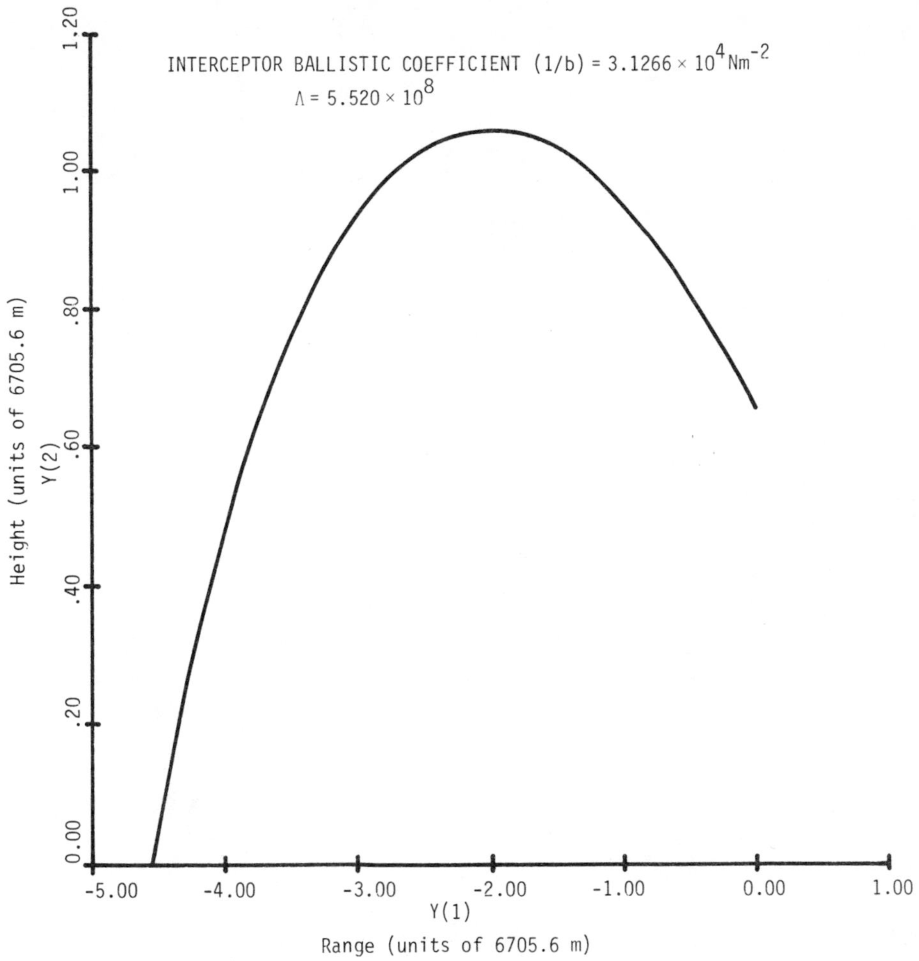

Figure 12 Schneider-Reddy Example. (Reprinted from NASA Contractor Report 3167, August, 1979.)

derivatives of functions. This can be a severe limitation in problems where the functions are not given in an analytic form but are given only as tabular data. Derivatives tend to be sensitive to errors in the tabular data and to the type of interpolation method used on the data. In addition, it may happen that no number of higher order corrections, no matter how accurate, gives a good approximation to the desired exact solution. In this case, the perturbation parameter in the problem is too "large" in some sense. Indeed, experience tends to show that a first order correction is sufficient in those cases in which an asymptotic approximation is going to be successful at all.

Other problems arise when we try to approximate nonlinear problems. Nonlinearity can introduce serious difficulties even for computing the lowest order approximation. Chapter 5 of O'Malley [33] contains an excellent discussion of examples of difficulties encountered in approximating nonlinear boundary value problems by the methods of singular perturbation theory. Similar problems would exist for other asymptotic methods. One problem that seems to occur often in trajectory optimization problems with bounded controls is that the boundary layers are not exponentially asymptotically stable as required in the usual asymptotic approximation theory of Tichonov and Vasileva for two-point boundary value problems (see Wasow [34]). In such cases the boundary layer region is actually finite and the boundary layer solution does not have the stable exponential behavior required by the conventional theory. Such a case occurs in the aircraft problem discussed in Washburn, Mehra, and Sajan [35] and Mehra, et al. [23] due to the presence of singular arcs. A simpler instance of this situation is illustrated in the example below.

Of course, the significance of the above-mentioned difficulties depends on the particular problem one is trying to solve, and generally, the success of an asymptotic method can be determined only by comparison of the asymptotic approximation with experiment or with a more exact solution computed by a less efficient method. We are interested in the off-line numerical solution of complex trajectory optimization problems, such as aircraft trajectory problems, in which the difficulties mentioned above are likely to be important issues. In particular, we have sought to find a numerical method which relies less on analytic expressions for functions and requires less mathematical analysis than the usual asymptotic methods, but is more efficient than the usual numerical optimization algorithms, which do not take account of asymptotic behavior. We believe that a very useful method with these properties can be obtained by a simple combination of asymptotic methods with numerical continuation methods. In the remainder of this section we will discuss a simple example to indicate how this procedure can be carried out. The reader should refer to Washburn and Mehra [36] for further details.

As a simple example of asymptotic continuation, consider Zermelo's problem (see Bryson and Ho [37]) of steering a boat in minimum time from one point to another when speed is constant and the rate of heading angle change is bounded. Mathematically, we have the system of differential equations

$$\dot{x} = \cos\sigma \quad (49)$$

$$\dot{y} = \sin\sigma \quad (50)$$

$$\sigma = u \quad (51)$$

where σ is the heading angle and

$$|u| \leqq 1 \quad (52)$$

Moreover, x and y must satisfy the boundary conditions

$$x(0) = x_0,\ y(0) = y_0 \quad (53)$$

$$\sigma(0) = \sigma_0 \quad (54)$$

$$x(t_f) = x_f,\ y(t_f) = y_f \quad (55)$$

The terminal time t_f is to be minimized.

The optimal solution of this problem is easy to determine and we can see clearly how an asymptotic continuation method behaves in this case. The optimal solution consists of an initial turn with $u = \pm 1$, followed by a straight path to the target with $u = 0$ (see Figure 13). To solve the problem using asymptotic continuation, we imbed (49-55) in a family of problems defined by changing (51) to

$$\lambda\dot{\sigma} = u \quad (56)$$

and by changing (54) to

$$\sigma(0) = (1 - \lambda)\sigma^* + \lambda\sigma_0 \quad (57)$$

where σ^* is given by

$$\tan\sigma^* = \frac{y_f - y_0}{x_f - x_0} \quad (58)$$

This is the first approach to asymptotic continuation described in Washburn and Mehra [36] and the solution trajectories in the x,y plane for λ between 0 and 1 are shown in Figure 14 for a particular case of boundary conditions.

It is easy in this case to see why it is necessary to continue the initial conditions as well as the differential equation if we want to start the continuation from the reduced order solution. In this example, the reduced order solution is just the $u = 0$, straight line trajectory between (x_0,y_0) and (x_f,y_f). However, if we do not replace (54) by (57), then a corrector (such as a gradient method) starting with the reduced

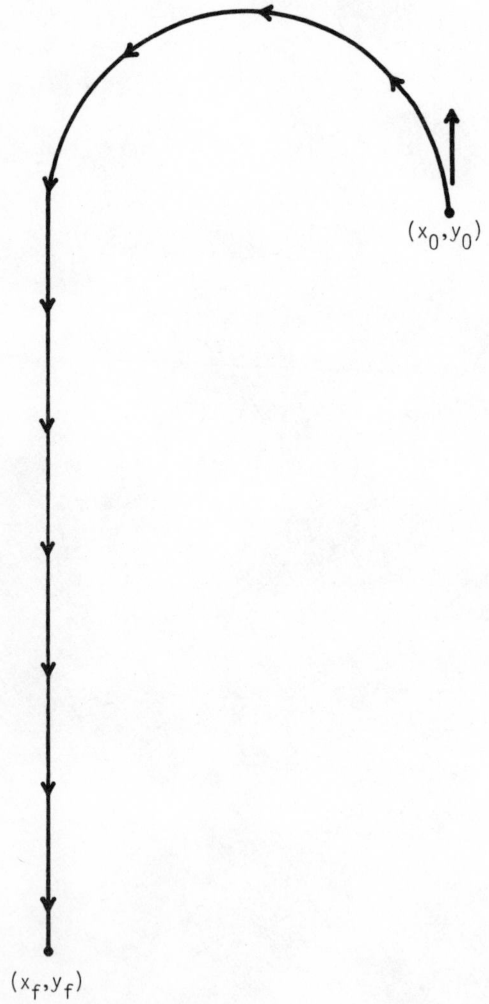

Figure 13 Optimal Trajectory in (x,y) Space for Zermelo's Problem. (Courtesy NASA.)

order solution gives as its first correction a straight line from (x_0, y_0) with angle σ_0 to the x-axis (see Figure 15). Thus, there is a big error in the terminal condition and the corrector must go through several iterations just to compute its first step in the continuation. This makes the method very inefficient and cancels any advantage of using the continuation method.

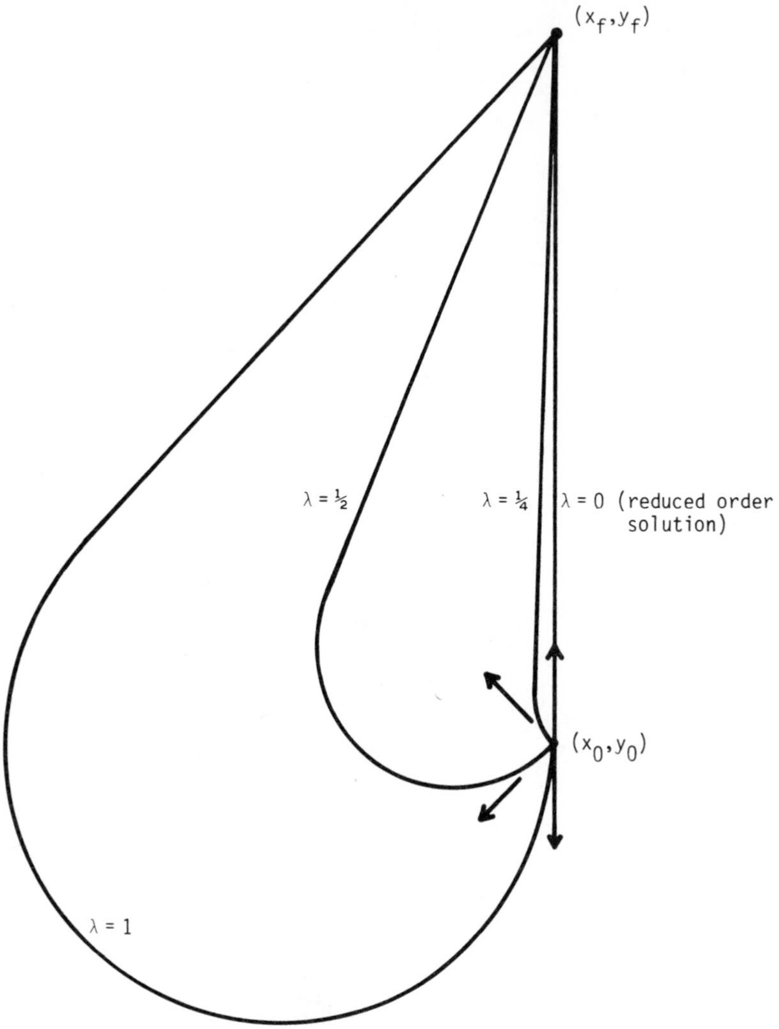

Figure 14 Continuation of Optimal Trajectory from $\lambda = 0$ to $\lambda = 1$. (Courtesy NASA.)

Note that the boundary layer equation

$$\frac{d\sigma}{d\tau} = u$$

where $\tau = t/\lambda$, is not exponentially asymptotically stable and one cannot use the currently developed asymptotic techniques to compute a first-order approximation with boundary layers.

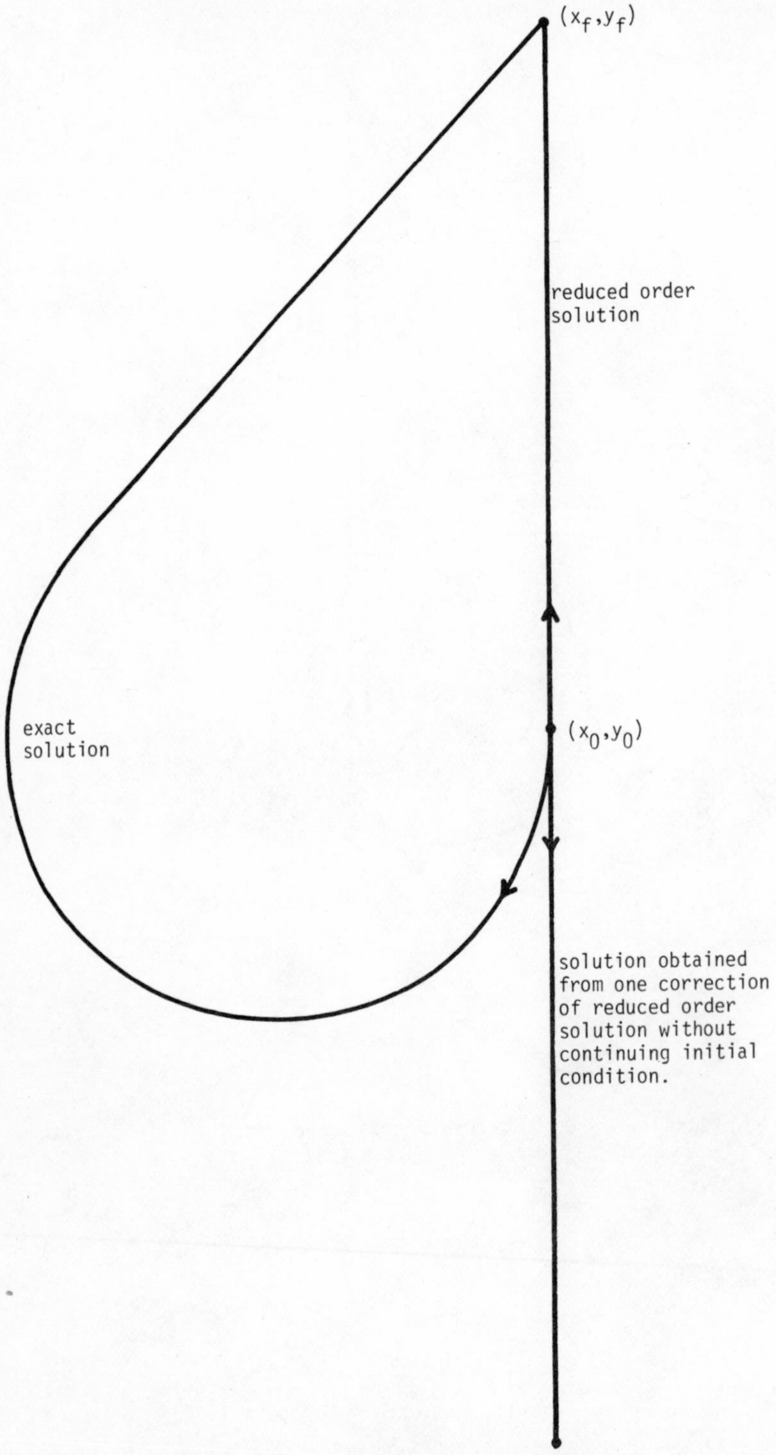

Figure 15 Necessity of Continuing the Boundary Conditions in Asymptotic Continuation Method. (Courtesy NASA.)

IV. CONCLUSIONS

Conceptually, one may view continuation methods as the intentional perturbation of the data of a problem in order to change a simple problem and its solution by continuous transformation into the desired problem and solution one is seeking. This conceptual procedure, once implemented as an algorithm, has successfully solved some difficult, nonlinear problems of engineering and shows promise for solving several other problems. Wider application of the method depends on further research to help understand the ideal theoretical and numerical behavior one can expect in practice. In particular, we identify three problems for further research whose solution would greatly aid applications. (i) As mentioned in Section II, there is hope that one may be able to compute all of the multiple solutions of a nonlinear problem by intentionally choosing continuation parameters that lead to bifurcation, as in Figure 4. However, there are needed general results that indicate when this is possible. Although Garcia and Zangwill [14] give some results in this direction, their conditions seem restrictive. Perhaps differential topology, applied along the lines of the homotopy methods described in Alexander [12], could provide a very general result. (ii) Some of the most appealing applications of continuation methods lie in the area of trajectory optimization. However, a much better understanding of possible bifurcation behavior is required in these cases. Current work emphasizes what can be generalized from the finite dimensional and linear cases. More work is needed on the infinite dimensional nonlinear cases that arise in trajectory optimization problems. (iii) Better understanding of the numerical aspects of continuation algorithms is needed, especially in cases where the algorithms are applied to large dimensional problems as in trajectory optimization or to problems exhibiting bifurcations. For example, bifurcations are indicated by the singularity of certain matrices. Given finite computational precision, one is likely to find that matrices are not singular but near singular. Is there a quantitative measure, depending on computational accuracy, that indicates how near to singularity a matrix must be for a bifurcation to be likely?

Acknowledgments

The authors gratefully acknowledge the support of the Department of Energy (contract ET-78-C-01-3389) and the Office of Naval Research (contract N00014-76-C-0724) in this research.

REFERENCES

1. C. M. Bender and S. A. Orszag, *Advanced Mathematical Methods for Scientists and Engineers*, New York: McGraw-Hill, 1978.
2. E. Lahaye, Une méthode de résolution d'une catégorie d'équations transcendantes, *C. R. Acad. Sci. Paris 198*(1934), 1840-1842.
3. D. Davidenko, On a new method of numerically integrating a system of nonlinear equations, *Dokl. Akad. Nauk. SSSR 88*(1953), 601-604 (in Russian).
4. W. C. Rheinboldt, An adaptive continuation process for solving systems of nonlinear equations, Computer Sciences Technical Report TR-393, University of Maryland, 1975.
5. H. Wacker, E. Zarzer, and W. Zulehner, Optimal stepsize control for the globalized Newton method, in *Continuation Methods* (H. Wacker, ed.), New York: Academic Press, 1978.
6. C. Den Heijer, *The Numerical Solution of Nonlinear Operator Equations by Imbedding Methods*, Amsterdam: Mathematisch Centrum, 1979.
7. H. B. Keller, Numerical solution of bifurcation and nonlinear eigenvalue problems, in *Applications of Bifurcation Theory* (P. H. Rabinowitz, ed.), New York: Academic Press, 1977.
8. M. Kubicek, Algorithm 502, dependence of solution of nonlinear systems on a parameter, *ACM-TOMS 2*(1976), 98-107.
9. M. G. Crandall and P. H. Rabinowitz, Bifurcation from simple eigenvalues, *J. Functional Anal. 8*(1971), 321-340.
10. W. C. Rheinboldt, Numerical methods for a class of finite dimensional bifurcation problems, *SIAM J. Numer. Anal. 15*(1978), 1-11.
11. B. J. Matkowsky and E. L. Reiss, Singular perturbations of bifurcations, *SIAM J. Appl. Math. 33*(1977), 230-255.
12. J. C. Alexander, The topological theory of an embedding method, in *Continuation Methods* (H. Wacker, ed.), New York: Academic Press, 1978.
13. L. T. Watson, Solving the nonlinear complementarity problem by a homotopy method, *SIAM J. Control and Optimization 17*(1979), 36-46.
14. C. B. Garcia and W. I. Zangwill, Global continuation methods for finding all solutions to polynomial systems of equations in n variables, Report 7755, Center for Mathematical Studies in Business and Economics, 1977.
15. C. B. Garcia and W. I. Zangwill, Determining all solutions to certain systems of nonlinear equations, *Math. of O.R. 4*(1979), 1-14.
16. J. M. Ortega and W. C. Rheinboldt, *Iterative Solution of Nonlinear Equations in Several Variables*, New York: Academic Press, 1970.
17. E. Wasserstrom, Numerical solutions by the continuation method, *SIAM Review 15*(1973), 89-119.
18. H. Wacker, A summary of the developments on imbedding methods, in *Continuation Methods* (H. Wacker, ed.), New York: Academic Press, 1978.
19. S. Roberts and J. Shipman, *Two Point Boundary Value Problems: Shooting Methods*, New York: American Elsevier, 1972.

20. M. Kubicek and V. Hlavacek, Solution of nonlinear boundary value problems--Va., a novel method: general parameter mapping (GPM), *Chem. Eng. Sci. 27*(1972), 743-750.

21. R. C. Montgomery, On the solution of optimal control problems by imbedding the terminal conditions, 4th Asilomar Conference on Circuities and Systems, November 18-20, 1970.

22. M. Jamshidi, On the embedding solution of a class of optimal control problems, *Automatica 8*(1972), 637-640.

23. R. K. Mehra, R. B. Washburn, S. Sajan, and J. V. Carroll, A study of the application of singular perturbation theory, NASA Contractor Report 3167, 1979.

24. R. K. Mehra, W. C. Kessel, and J. V. Carroll, Global stability and control analysis of aircraft at high angles-of-attack, ONR Report ONR-CR215-248-1, 1977.

25. J. E. Marsden and M. McCracken, *The Hopf Bifurcation and Its Applications*, New York: Springer-Verlag, 1976.

26. H. Schneider and P. B. Reddy, A new optimization technique for solving nonlinear two-point boundary value optimal control problems with application to variable vector thrust atmospheric interceptor guidance, AIAA Paper 74-827, 1974.

27. P. J. Orava and P. A. J. Lautala, Back and forth shooting method for solving two-point boundary value problems, *J. Optimization Theory and Appl. 18*(1976), 485-498.

28. P. J. Orava and P. A. J. Lautala, Interval length continuation method for solving two-point boundary value problems, *J. Optimization Theory Appl. 23*(1977), 217-227.

29. A. H. Nayfeh, *Perturbation Methods*, New York: Wiley, 1974.

30. P. V. Kokotovic, R. E. O'Malley, and P. Sannuti, Singular perturbations and order reduction in control theory--an overview, *Automatica 12*(1976), 123-132.

31. M. D. Ardema, Solution of the minimum time-to-climb by matched asymptotic expansions, *AIAA J. 14*(1976), 843-850.

32. R. V. Ramnath and P. Sinha, Dynamics of the space shuttle during entry into the earth's atmosphere, *AIAA J. 13*(1975), 337-342.

33. R. E. O'Malley, *Introduction to Singular Perturbations*, New York: Academic Press, 1974.

34. W. Wasow, *Asymptotic Expansions for Ordinary Differential Equations*, New York: Krieger, 1976.

35. R. B. Washburn, R. K. Mehra, and S. Sajan, Application of singular perturbation techniques (SPT) and continuation methods for on-line aircraft trajectory optimization, *Proc. Conference on Decision and Control* (1978), 983-990.

36. R. B. Washburn and R. K. Mehra, Asymptotic continuation method for trajectory optimization, *Proc. Joint Automatic Control Conference* (1979), 617-621.

37. A. E. Bryson and Y.-C. Ho, *Applied Optimal Control*, Washington, D.C.: Hemisphere, 1975.

Chapter 12 ESTIMATION OF HYDROGEOLOGIC PARAMETERS FOR OPTIMAL GROUNDWATER BASIN DEVELOPMENT

JOHN W. LABADIE and DUANE R. HAMPTON / Colorado State University, Fort Collins, Colorado

I. ABSTRACT

Hydrogeologists are in need of a computerized tool for aiding them in groundwater basin development decisions. Increasing demand for water supply for municipal, industrial (including energy development), and agricultural uses, coupled with the small number of good remaining sites for large surface water impoundments, has focused attention on further exploitation of our extensive groundwater resources. There is, however, a dearth of data on basin characteristics and hydrogeologic parameters upon which to base sound developmental decisions. A methodology has been developed as an adjunct to the experience and intuition of the hydrogeologist. Three categories of modeling technology are synthesized to produce the desired methodology: (1) groundwater flow prediction models; (2) the Kalman filter for incorporating uncertainties associated with sparse and inaccurate data and approximate flow prediction models, and for finding best minimum variance estimates of hydrogeologic parameters; and (3) dynamic programming, which is ideally suited to stochastic decision processes involving nonconvex objective functions and state-space constraints. Though the major contribution is in the synthesis of these elements, a technique for improved convergence of the extended Kalman filter is developed. Through use of the synthesized methodology, the hydrogeologist is provided least-cost development guidelines, under risk constraints, which can be updated in real-time as new information becomes available.

II. INTRODUCTION

Increasing demand for water supply for municipal, industrial, and agricultural uses, coupled with the fact that there are few remaining good sites for large surface water impoundments, has created an impetus toward greater development of our vast groundwater resources. Potable groundwater is available at varying depths throughout most of the U.S. According to Todd [1], total usable groundwater in storage is comparable to about 35 years of runoff; and yet only about 20 percent of total water usage comes from groundwater. Water stored underground in aquifers is often of high quality and can be consumed directly, while surface waters usually require treatment before use. Groundwater reservoirs incur insignificant evaporative losses, unlike surface water impoundments. Although pumping groundwater requires energy, energy development (oil shale or coal) requires groundwater in the water poor West. Thus, groundwater may be extremely important not only for Western irrigated agriculture, industry, and municipalities, but to the national effort to attain energy self-sufficiency.

The optimal development of a groundwater basin is conceptually based on a sequential process of drilling wells to obtain both water and hydrogeologic information. This information, in turn, is used as input for augmentation of a well network for supplying additional water. Given the usual budgetary constraints, an optimum design for the network would be one where initial wells are successful and can be incorporated into the final configuration of water supply wells. Therefore, wells must be drilled not only to obtain water, but to obtain more information that can serve as a basis for obtaining more water.

Groundwater systems are characterized by a large number of parameters (i.e., spatially distributed storage coefficients and transmissivities). Direct determination of parameters through in situ measurement or laboratory tests of samples is not economical, and often not even meaningful, since the samples tend to reflect highly localized conditions.

An indirect method for inferring the system parameters is through study of well water level response to given excitations (pumping tests).

This customary way of estimating parameters depends on several simplifications about the system and gives a lumped measure of the system characteristics at the location of the test. Because of the inherent uncertainty and nonhomogeneity of natural hydrologic systems, there is an underlying problem of parameter identifiability. In spite of this, it has been found that these lumped estimates produce sufficiently accurate predictions of basin behavior. Therefore, the well we use to obtain water is also used to obtain data on basin characteristics.

Groundwater basin development decisions with respect to where and how many wells of what capacities should be drilled are largely based on the experience and intuition of the hydrogeologist. Significant advances in groundwater basin modeling (as summarized by Bachmat, et al. [2]) have taken place over the last 10 to 15 years. The actual application of this technology to groundwater basin development is virtually nonexistent. This is perhaps due, in part, to the fact that current sophisticated models require an enormous amount of data for calibration, verification, and prediction. Such data are simply not available in newly developed or underdeveloped basins.

Our purpose here is to present a methodology which could conceivably be a useful tool for the hydrogeologist involved in groundwater basin development. Three categories of modeling technology are synthesized to produce the desired methodology:

1. Groundwater flow prediction models to define at least some causality between groundwater basin development decisions and basin response, in which the latter is usually measured in terms of basin water level or piezometric head.
2. The Kalman filter, which represents an effective means of incorporating into the predictive modeling the uncertainties associated with sparse data, inaccurate data, and predictive models which can only approximate the behavior of a complex system such as a groundwater basin.
3. Dynamic programming, which is ideally suited to stochastic sequential decision processes involving nonconvex objective functions and state-space constraints.

We admit to making little contribution in these individual categories, except to propose a successive-approximations procedure in the Kalman filter algorithm which can potentially aid convergence, and to develop a new dynamic programming algorithm designed to deal with the large number of state variables needed to model groundwater basins. The important contribution, however, is in the synthesis. Details of the dynamic programming algorithm will not be presented here. The interested reader is referred to [3].

It is interesting that optimization theory has developed along the two major lines of control theory and mathematical programming, though there is obviously a basic unifying theory which several authors have identified. Kalman filtering has come out of the former, and there are many operations research specialists who are unfamiliar with its attractive properties. We have attempted to bridge that gap by showing how dynamic programming and Kalman filtering can be explicitly combined together in a rational methodology for making decisions under uncertainty (or risk). This work is a first step only, and much remains before a practical tool can be developed that can aid the hydrogeologist in the field.

A review of the important literature related to groundwater basin development follows. Since we approach this problem from the viewpoint of a groundwater hydrologist or hydrogeologist, we may have missed some of the important recent contributions to this area from the petroleum industry. Groundwater flow prediction models are discussed next. This is a vast area, however, and we only touch on the subject sufficiently to show how these models can be incorporated into a decision framework in an uncertain environment. Our discussion of the application of Kalman filtering draws heavily on the pioneering work of Kitanidis [4] and McLaughlin [5]. Some computational results are presented, but they are incomplete. They are useful primarily to show the computational feasibility of the methodology, rather than in any way *proving* its validity in improving groundwater basin development decisions. This is left for future work. It is believed, with inclusion of appropriate system state models, that the basic methodology presented here could be applied to petroleum exploration, selection of sites for underground depositories of nuclear waste, watershed management, rangeland management, optimal irrigation scheduling, pollution control, and water quality management.

III. DESCRIPTION OF THE GROUNDWATER MODEL

To manage optimally the groundwater resource in a basin, the response of the system to natural inputs and human interference must be predictable. In an effort to quantify that response, mathematical equations have been formulated which describe the system performance. In most practical problems, these equations can only be solved by numerical techniques implemented on the digital computer.

No attempt will be made herein to formally derive the basic groundwater flow equation. The reader is referred to Freeze and Cherry [6] for this. Several definitions will be given, however, in order to foster an understanding of its implications. The groundwater flow equation describes flow (usually of water) through *porous media*. Any material with interconnected interstices large enough to permit flow may be considered porous media.

The *state* of a groundwater basin is usually defined in terms of the *piezometric head* h(x,y,t), where (x,y) are spatial coordinates and t is time. These arguments will be deleted henceforth for notational convenience. The piezometric head is the sum of the gravitational potential energy due to vertical water level position relative to an arbitrary datum and any additional pressure head $(P - P_0)/\rho g$; i.e., $h = z + (P - P_0)/\rho g$. The reference pressure P_0 is usually selected to be atmospheric pressure, which is assigned a value of zero; ρ is the density of the fluid and g is gravitational acceleration.

A. Important Parameters

The manner in which fluids flow through porous media, in a macroscopic sense, is described by Darcy's Law:

$$Q = -KA \frac{dh}{dx}$$

where Q is the volumetric flow rate, x is the direction of flow, A is the cross-sectional area perpendicular to x through which flow occurs, and K is a proportionality constant [which actually varies spatially; i.e., $K \equiv K(x,y)$] called *hydraulic conductivity* with units of length per time. The minus sign is present to ensure that flow occurs in the direction of decreasing head, or potential energy. Note that Q is directly proportional to K. Porous media which have a higher hydraulic conductivity are more likely to provide a larger volumetric flow with all other factors being equal.

The groundwater resource is contained in *aquifers*. An aquifer is composed of porous media, such as layers of rock, sand or silt, which store and transmit water in significant quantities. There are two major types of aquifers, as seen in Figure 1. A *confined aquifer* is one in which water is stored under pressure. This occurs because it is confined above and below by relatively impermeable strata, such as clay layers or bedrock. *Unconfined aquifers* have no confining upper layer, so that the water level (or *water table*) is essentially at atmospheric pressure. There are other types of complex aquifer conditions that can occur.

The physical parameters used to describe a groundwater basin differ, depending upon whether an aquifer is confined or unconfined. The saturated thickness of an unconfined aquifer is h, the height of the water table.

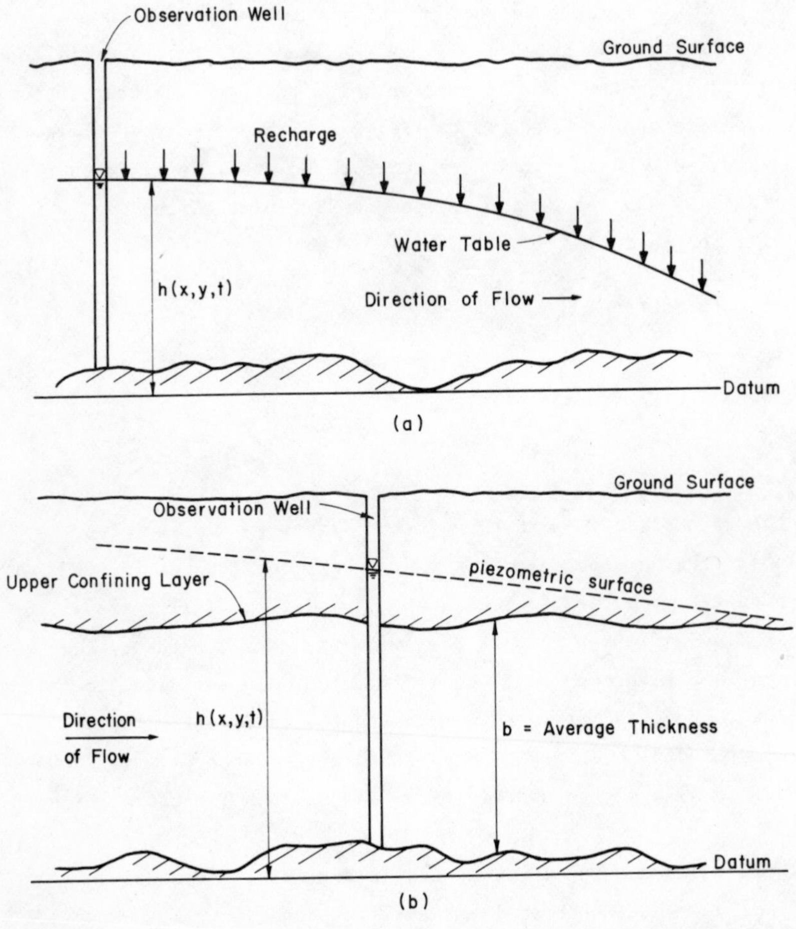

Figure 1 (a) Illustration of an unconfined aquifer; (b) confined aquifer.

The saturated thickness of a confined aquifer, b, is a constant equal to the average thickness of the aquifer per unit width. The product of K and saturated thickness is the *transmissivity*, T. This is constant with respect to head h in a confined aquifer, whereas it varies linearly with h in an unconfined aquifer. Parameter T, however, will vary not only with spatial coordinates (x,y), but also with the direction of flow. The *storage coefficient* S is a dimensionless parameter that, for a confined aquifer, can be interpreted as the volume of water released from a column of unit area and height b, per unit decline in pressure head. For unconfined aquifers, S is actually *specific yield* S_y, which can vary between 0.01 to around 0.40. It has essentially the same meaning as S, but is not dependent on b. The storage coefficient for confined aquifers is much smaller (from 10^{-5} to about 10^{-2}) than that for unconfined aquifers. That is, the head will decline more rapidly in a confined aquifer than in an unconfined aquifer when both are pumped at the same volumetric flow rate.

As pointed out by McLaughlin [5], parameter S is generally not as important as hydraulic conductivity K in predicting groundwater flow. This is evident from Darcy's law, which states that groundwater flow rate is directly proportional to K. Since transmissivity T is in turn directly proportional to K, we will focus on this as the major parameter of interest in this study. It is, however, difficult to obtain accurate estimates of T in the field without actually pumping wells in the basin and noting the resulting decline in head at observation wells. The dilemma is that good estimates of T are needed to decide where to drill new wells. If the basin is newly developed, there may be an insufficient data base for determining T.

B. Groundwater Flow Equation

The basic two-dimensional equation governing unsteady groundwater flow is derived from Darcy's law, conservation of mass, and certain important assumptions [6]:

$$\frac{\partial}{\partial x}\left(T_x \frac{\partial h}{\partial x}\right) + \frac{\partial}{\partial y}\left(T_y \frac{\partial h}{\partial y}\right) = S\frac{\partial h}{\partial t} + q \qquad (1)$$

This equation ignores vertical flow, and parameters T_x, T_y, and S are vertically averaged. Transmissivities are subscripted x and y to allow for the possibility of different transmissivities at a given point, depending on the direction of flow. Aquifers with this property are called *anisotropic*. If S and T vary spatially [i.e., $S \equiv S(x,y)$ and $T \equiv T(x,y)$], the aquifer is *nonhomogeneous*. More detailed, three-dimensional equations can be derived, but are considered too detailed for the purpose of this study.

Appropriate initial and boundary conditions must be specified with Equation (1):

1. $h(x,y,0) = h_0$ (i.e., a steady state water level prior to extensive pumping)
2. $h(x',y',t)$ specified for lakes and/or streams bounding the aquifer at locations (x',y')
3. $[dh(x'',y'',t)]/dn = 0$, where n is the direction normal to an impermeable boundary at locations (x'',y''). A water table of zero slope implies zero flow across the boundary.
4. Other boundary conditions, such as constant subsurface discharge (or recharge).

The term $q(x,y,t)$ in Equation (1) is called the source/sink term, and represents vertical recharge to or discharge from the aquifer. The units are volumetric flow rate per unit cross-sectional area of flow. It may include natural recharge, such as deep percolation from rainfall, but is generally dominated by pumpage at some specific location (x_i,y_i) at well i. For unconfined aquifers, $S = S_y$ and $T = Kh$, thereby rendering the equation nonlinear. Often, however, it is acceptable to assume that T is not appreciably affected by h if the depth of water saturated thickness of the aquifer is large in comparison with *drawdown*. By drawdown, we mean reduction in h from some initial steady state level h_0.

For nonhomogeneous, anisotropic aquifers with complex boundary conditions, Equation (1) cannot be solved analytically. An approximate numerical solution can be obtained by finite elements or finite differences. The latter approach is still the most popular among groundwater modelers, and involves dividing up the basin into homogeneous grids or *nodes*. Figure 2 gives an example discretization with nine interior nodes and 16 exterior nodes. The latter are used to represent boundary conditions. Term TX(K,L) represents transmissivity in the x-direction and TY(K,L) is transmissivity in the y-direction. The heads H(I), where I = (L - 1)3 + K, are assumed measured at the node points at the center of each grid. Obviously, accuracy is increased with more and smaller grids, but computational requirements increase accordingly.

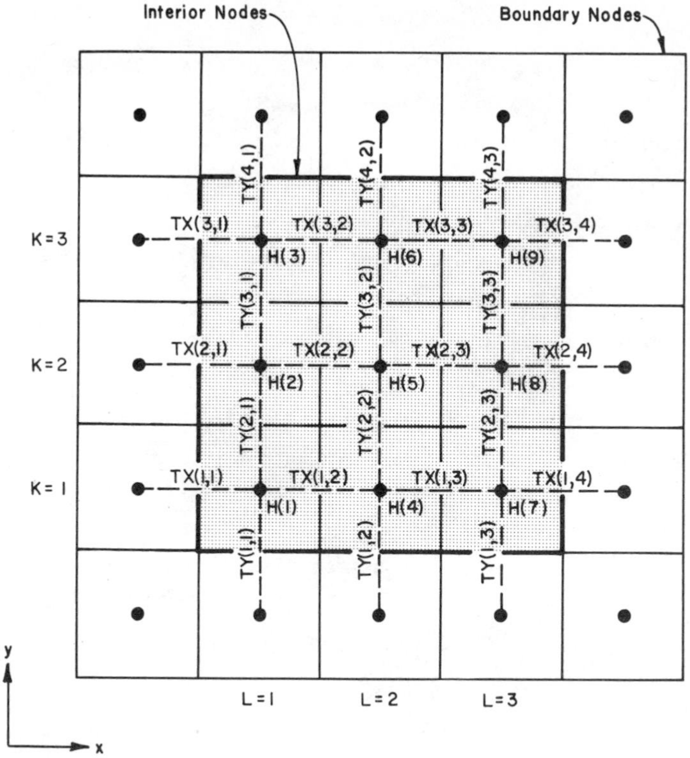

Figure 2 An example finite difference grid system for a groundwater basin.

There are several ways of performing the numerical differencing in order to approximate the derivatives on the left-hand side of Equation (1). Whatever the approach, Equation (1) can thus be approximated by a set of simultaneous first-order linear differential equations of the form:

$$\frac{d\underline{h}}{dt} = A\underline{h} + B\underline{u} + \underline{c}$$
$$\underline{h}(0) = \underline{h}_0 \tag{2}$$

It is possible to restrict the dimension of $\underline{h}$ to the number of interior nodes and incorporate boundary conditions in an appropriate selection of $\underline{u}$ and $\underline{c}$ for nodes bordering the boundaries. (See [7] for details.) Assuming there is a total of n nodes, the A-matrix is an n × n symmetric matrix whose structure depends upon the type of differencing used. It is composed of parameter vectors $\underline{T}$ and $\underline{S}$, whereas matrix B includes only $\underline{S}$. The constant vector $\underline{c}$ includes both $\underline{S}$ and $\underline{T}$. Controllable and uncontrollable discharge/recharge is separated by including natural vertical recharge/discharge in the constant term $\underline{c}$, and letting the vector $\underline{u}$ represent only controllable pumping or artificial recharge. The emphasis here is on pumping. Therefore, if wells exist at $m < n$ nodes, then matrix B is $n \times m$. If there are several wells in a particular node, we aggregate their discharge capacities. The vector $\underline{u}$ then is an m-dimensional decision vector in units of volumetric flow rate per unit area.

Field measurements of groundwater piezometric head levels are generally made at discrete time intervals (as long as a month or more). Well field development and management decisions are also made at discrete intervals. Therefore, Equation (2) should be approximated by difference equations. The techniques of linear system theory [8] are applicable here, where we analytically solve the simultaneous linear differential equations by assuming that decision vector $\underline{u}(t)$ is constant over *discrete* interval $[t, t + \Delta t)$. (Henceforth, we will regard t as a discrete index.) This results in the following set of algebraic difference equations:

$$\underline{h}_{t+1} = \phi(\underline{T})\underline{h}_t + \psi(\underline{T})\underline{u}_t + \Omega(\underline{T})\underline{c}_t \qquad \text{(for discrete } t = 1,\ldots,N) \qquad (3)$$

$$\underline{h}_1 = \text{given initial conditions}$$

Notice that all of the matrices are functions of transmissivity $\underline{T}$, which is the principal parameter we will be considering.

IV. REVIEW OF THE LITERATURE

The problem of optimal groundwater basin development when all aquifer parameters are assumed known has been approached from several directions. Analytical techniques, finite differences coupled with linear programming, common sense, and engineering "cookbook" approaches have all been used to determine the optimal locations and pumping rates of several wells.

Hantush [9] derived analytical solutions for optimum well spacing for various geometries, based on the cost of connecting pipelines and the cost of additional drawdown due to well interference.* The aquifer can be leaky or nonleaky and have variable thickness. The case of a water table aquifer having a sloping base is also considered. The assumptions include:

---

*By *well interference* we mean a situation where wells are sufficiently close together to produce drops in water level at a well due to pumping at adjacent wells. A group of interfering wells is called a *well field*.

(1) an aquifer of infinite areal extent, (2) fully penetrating wells, and (3) aquifer homogeneity.

Alley, Aguado, and Remson [10] developed a method of groundwater management for a confined aquifer under steady-state or transient two-dimensional flow using linear programming and finite differences. The solutions for the resulting linear programming problem are used to determine optimal well distributions and pumping rates. The aquifer parameters must be known, but transmissivity may vary. The wells are located in a square grid pattern.

A common sense approach is taken in [11]. According to this article, the most desirable arrangement in an extensive aquifer is to locate wells at equal spacings on the circumference of a circle. Or, where a source of recharge is known to exist nearby, the wells might be located in a semicircle or along a line roughly parallel to the source of recharge. Multiple wells are considered to be preferable in aquifers that are shallow, less permeable, or varied in transmissivity. The article claims that single well installations are better where the water-bearing formation is of considerable thickness and of sufficient permeability.

An example of a precalculated solution to the optimal well spacing problem is that developed by Mount [12]. Tables are given which allow computation of the drawdown at wells nearest the center of a rectangular array of wells and at the corner of the well field. The *optimal* pattern is then found by varying the number of rows and columns and observing the resulting drawdown at the center of the well field (where it will be greatest). One would choose the pattern that requires the fewest wells, without producing excessive drawdown. Of course, aquifer parameters and water requirements must be known, as well as the area of the well field.

The problem of optimal well field development when the aquifer parameters are unknown or uncertain has not been conclusively addressed as yet. Several related efforts directed toward differing aspects of the overall problem are available in the literature.

Pfannkuch and Labno [13] discuss the design of groundwater monitoring networks for pollution studies. Their explanation of the nature of the problem and use of systems analysis techniques to decompose it into phases can be applied to the more general problem of well field development. In both cases, the successful design and operation of a well field are based on a stepwise process of obtaining hydrogeologic information. This, in turn, is used as input for a more rational design of the network to meet project objectives. Outlines and flow diagrams for this step-by-step thought process for developing a network of wells are presented. No methodology for optimizing the network design is given, other than the overall planning strategy.

Bostock, Simpson, and Roefs [14] attempt to solve the well field design problem by considering a two-dimensional uniform grid of wells in an infinite aquifer. Each grid contains a pumping well, which is assumed to behave the same as all other wells. All aquifer properties are known and are homogeneous, with the exception of permeability. The estimated frequency of each grid permeability value is defined by a probability density function. A method of calculating costs due to well spacing and pumping lift is developed for a desired production rate from the entire well field. The sum of these two costs is minimized by computing the cost for several discrete values of well capacity (or well spacing) and selecting the lowest one. Additional assumptions associated with this approach include: (1) all wells have the same design, are equally spaced, and form a uniform grid; (2) the aquifer has infinite areal extent, uniform thickness, and is horizontal; (3) the initial water table is horizontal; and (4) the wells are fully penetrating and commence and continue pumping simultaneously at identical discharge rates.

The last assumption is quite limiting because well fields are not usually developed all at once in a basin where the parameters are uncertain. In such situations, drilling a single well is the first step taken. If it is productive, more wells may be drilled nearby.

Many other papers attack the so-called *inverse problem*, in which unknown aquifer parameters are calculated based on aquifer response to pumping. There is no attempt to find an optimal well field design. Chang and Yeh [15] introduce an algorithm for parameter identification in a two-dimensional confined aquifer. A least-squares objective function is employed and the identification problem is formulated as a quadratic programming problem. Their algorithm is shown to be effective in solving the large-scale inverse problem.

Wilson, et al. [7] propose, based on a master's thesis by Kitanidis [4], the use of an extended Kalman filter to estimate the state and parameters of a groundwater basin. This method accounts for model and measurement error, as well as parameter uncertainty. Prior information about the parameter is utilized, as well as information extracted from the input-output measurements of the system. The groundwater flow equation is transformed into a set of simultaneous system state prediction equations using state-space concepts. An exponentially stable time integration scheme is used. Second moment analysis gives rise to predictive equations for the mean and variance of the state (piezometric head) and the parameters (transmissivity). The extended Kalman filter, which gives the minimum variance estimate for the state and parameters of the system, is formulated.

The assumptions made by Wilson, et al. [7] are the following: (1) the storage coefficient is constant and uniform; (2) the initial head distribution is known and is error-free; (3) the aquifer behaves linearly; (4) the transmissivity is constant, but unknown; (5) some initial estimates of transmissivity and the uncertainty of those estimates are available; (6) the boundary conditions and pumping rates are known with zero error; and (7) the set of grids at which the state is measured is constant.

Lenton, Schaake, and Rodriguez-Iturbe [16] propose the use of Bayesian techniques for parameter estimation with limited data. Such techniques improve the available limited hydrologic data by taking into account all the information coming from other sources, both objective and subjective. These authors claim that classical methods of estimation are defective since they produce values that are independent of the economic consequences of erroneous estimates. In their approach, the expected opportunity losses caused by an erroneous parameter estimate are assessed and then minimized as a criterion for estimation. Application is made to parameter estimation for a first order autoregressive model.

While Gates and Kisiel [17] do not attempt to solve the inverse problem, they do try to specify which additional data could most improve a computer model of the Tucson basin. The model variables for which further data are considered include: the storage coefficient, transmissivity, initial head distribution, discharge, and recharge. The *worth of data* is evaluated in terms of the expected reduction in error in predicted water levels associated with the collection of more data for various parameters and inputs over the basin. They conclude that the Tucson basin model could be improved most by obtaining more data on discharge and recharge in areas where these variables are large, and on transmissivity where it is uncertain. More data on initial water levels and storage coefficient would be less helpful according to these authors.

A review paper by McLaughlin [5] describes several applications of Kalman filtering concepts to groundwater basin management. Most of the theoretical approaches suffer from the sparsity of groundwater data and statistical information such as error covariances. McLaughlin also mentions that a statistical approach to defining spatially distributed piezometric levels (when actual data are collected at a small number of discrete locations) called *kriging* (Delhomme [18]) is actually a specialized application of Kalman filtering. Gelhar [19], on the other hand, has focused on the statistical analysis of aquifer parameters such as hydraulic conductivity as a means of developing better predictions of groundwater flow.

V. KALMAN FILTER ALGORITHM

The Kalman filter [20] has been widely applied by electrical engineers and control system analysts for years. As mentioned previously, operations research specialists who have looked at optimization from the mathematical programming viewpoint have only recently begun to study the Kalman filter and note its attractive features. The last five years have seen its application in civil engineering, particularly water resources engineering [21]. Most of the applications have been in pollution control and water quality management. The first applications in groundwater modeling were by McLaughlin [5] and Kitanidis [4].

A. System State Prediction Model

The Kalman filter is no more than a *recipe* for combining two independent estimates of a state vector to provide a *best* (minimum variance) estimate of the system state. The two independent estimates of the system are given by: (1) a process model or *system state prediction model*, based on a priori understanding of the prototype system; and (2) direct *measurements* of some or all of the state variables. Estimates from the prediction model contain uncertainty due to inaccuracies and approximations associated with the model as an attempt to simulate the behavior of a complex system. Similarly, the measurements contain sampling and analytical errors. The filter combines the model and data estimates by weighting them in such a way that the uncertainty of the final estimate is less than the uncertainty associated with either independent estimate individually; that is, a minimum variance estimate is obtained. Output from the filter consists of a new improved estimate of the state of the system and the variances and covariances associated with that estimate [22].

Young [23] has shown that the Kalman filter is related to recursive least-squares estimation of the parameters of linear models. The Kalman filter is only applicable to linear systems, or systems that behave approximately linearly in the neighborhood of a nominal state trajectory.

In the *extended* Kalman filter, as applied by Kitanidis [4], parameters of the prediction model are added to the vector of state variables. The resulting prediction model is nonlinear, which requires linearization around the latest estimate of the augmented state vector. The piezometric heads $\underline{h}$ and transmissivities $\underline{T}$ are regarded as random variables. Uncertainties in the storage coefficient estimates $\underline{S}$, boundary conditions, and natural recharge/discharge are not considered. Our subsequent formulation will assume the same. Decision vector $\underline{u}_t$ is regarded as completely controllable and $\underline{c}_t$ is deterministic. It should be noted that according to Labadie [24], uncertainties in boundary conditions can possibly be incorporated into the parameter estimates. That is, if boundary conditions are not well known, final parameter estimates may not agree with actual field estimates, but rather reflect the effects on water level response of boundary conditions and other complex basin characteristics. For this reason, Labadie [24] refers to them as *surrogate parameters*, since their primary purpose is providing an acceptable excitation-response (or input-output) mechanism.

According to Kitanidis [4], the state-space is augmented to include the transmissivities as follows:

$$\underline{x}_{t+1} = \begin{bmatrix} \underline{h}_{t+1} \\ \underline{T}_{t+1} \end{bmatrix} = \begin{bmatrix} \phi(\underline{T}_t)\underline{h}_t + \psi(\underline{T}_t)\underline{u}_t + \Omega(\underline{T}_t)\underline{c}_t \\ \underline{T}_t \end{bmatrix} + \begin{bmatrix} \underline{\varepsilon}_t \\ \underline{0} \end{bmatrix} \tag{4}$$

(for $t = 1,\ldots,N$)

where a residual error term $\underline{\varepsilon}_t$ is added to account for model error, or *noise*.

The following assumptions are associated with the noise term:

(i) $E\{\underline{\varepsilon}_t\} = \underline{0}$ for all t

(ii) $E\{\underline{\varepsilon}_t\underline{\varepsilon}_\tau'\} = Q\delta_{t\tau}$ for all t,τ

in which $\delta_{t\tau}$ is the Kronecker delta and E is the expectation operator. Therefore, we assume the error term is independently distributed with known covariance matrix Q. The prediction model (4) must, of course, require that transmissivities not change from period to period, which accounts for the lack of a noise term for $\underline{T}$. When observations are taken, however, estimates of $\underline{T}$ will be allowed to change.

Since Equations (4) are nonlinear, the usual approach, and the one taken by Kitanidis [4], is to linearize them around current estimates of $\underline{h}$ and $\underline{T}$ using a truncated Taylor series. This is a tedious procedure which the authors have observed to display poor convergence characteristics. Initial estimates of transmissivity covariances are required. This is difficult since there is generally little information on which to base these estimates. Nelson and Stear [25] also confirm the divergent tendencies of the extended Kalman filter. We use a somewhat different approach here, as suggested by Graupe [26]. A *successive approximations* approach is employed, wherein we linearize (4) by assuming initial estimates $\hat{\underline{T}}_t$ are given *before* we find $E\{\underline{h}_{t+1}\}$. We will subsequently show how these estimates are updated. Assuming current estimates for the ℓth iteration $\underline{T}_t^{(\ell)}$ are given, then

$$\underline{h}_{t+1} = \phi\left(\hat{\underline{T}}_t^{(\ell)}\right)\underline{h}_t + \psi\left(\hat{\underline{T}}_t^{(\ell)}\right)\underline{u}_t + \Omega\left(\hat{\underline{T}}_t^{(\ell)}\right)\underline{c}_t + \underline{\varepsilon}_t \tag{5}$$

$$E\{\underline{h}_{t+1}\} = \hat{\underline{h}}_{t+1} = \phi\left(\hat{\underline{T}}_t^{(\ell)}\right)\hat{\underline{h}}_t + \psi\left(\hat{\underline{T}}_t^{(\ell)}\right)\underline{u}_t + \Omega\left(\hat{\underline{T}}_t^{(\ell)}\right)\underline{c}_t \tag{6}$$

where $E\{\underline{h}_{t+1}\}$ is conditioned on $\hat{\underline{T}}_t^{(\ell)}$. Since we are assuming, in this iteration, that $\hat{\underline{T}}_t^{(\ell)}$ does not change [that is, the $\underline{T}$ are temporarily regarded as given values, $\hat{\underline{T}}^{(\ell)}$], it follows that the conditional error covariance is

$$\begin{aligned} E\{[\underline{h}_{t+1} - \hat{\underline{h}}_{t+1}][\underline{h}_{t+1} - \hat{\underline{h}}_{t+1}]'\} &= P_{h(t+1)} \\ &= \phi\left(\hat{\underline{T}}_t^{(\ell)}\right)P_{h(t)}\phi\left(\hat{\underline{T}}_t^{(\ell)}\right)' + Q \end{aligned} \tag{7}$$

We also assume that estimates of the mean and error covariance for initial heads are given:

(i) $E\{\underline{h}_1\} = \hat{\underline{h}}_1$

(ii) $E\{(\underline{h}_1 - \hat{\underline{h}}_1)(\underline{h}_1 - \hat{\underline{h}}_1{}')\} = P_{h(1)}$

B. Observation Model

In addition to the predictive capability of our model, we have the opportunity of successively taking direct measurements on the piezometric heads over the groundwater basin. The observations are assumed to be made at the end of period t, and are denoted by $\underline{z}_{t+1}$. The observation model is

$$\underline{z}_{t+1} = C\underline{h}_{t+1} + \underline{\eta}_{t+1} \tag{8}$$

in which the matrix C is composed of zeroes and ones to denote in which nodes measurements are taken. Notice that we do not include transmissivities here since they cannot be directly measured in the field. The vector $\underline{h}_{t+1}$ represents the true, but unknown heads, and $\underline{\eta}_{t+1}$ is a noise term of the same dimension as $\underline{z}_{t+1}$, assumed also to be independently distributed with zero mean and known covariance matrix R.

With inclusion of these observations, it can be proved by several methods (see [22]) that the minimum variance estimate of $\underline{h}_{t+1}$ is

$$\hat{\underline{h}}_{t+1|t+1} = \hat{\underline{h}}_{t+1} + G_h[\underline{z}_{t+1} - \bar{C}\hat{\underline{h}}_{t+1}] \tag{9}$$

with error covariance

$$P_{h(t+1|t+1)} = P_{h(t+1)} - G_h C P_{h(t+1)} \tag{10}$$

where

$$G_h = P_{h(t+1)} C'[R + CP_{h(t+1)} C']^{-1} \tag{11}$$

is called the *Kalman gain*. The estimates $\hat{\underline{h}}_{t+1}$ are actually obtained from the system state prediction model [Equation (6)]. We show a conditioning on t+1 for the estimates $\hat{\underline{h}}_{t+1|t+1}$ to indicate that these are updated estimates based on observations taken at the end of period t. Notice from Equation (9) that the magnitudes of the gain components influence the degree of change in the updated state estimate. However, if the observations exactly coincide with current estimates, the updated estimates do not change. Notice also that implied in Equation (9) is a reduction in variances and covariances as observations are taken, and hence a reduction in uncertainty as we learn more about the system.

Suppose we decide that the current best estimate of $\underline{T}$ at the end of period t is that which produces an output from the groundwater prediction model (6) which coincides (within a specified tolerance) with head estimates produced by Equation (9). In effect, we want to solve the following optimization problem:

$$\min_{\underline{T}} [\hat{\underline{h}}_{t+1|t+1} - \hat{\underline{h}}_{t+1}(T)]'[\hat{\underline{h}}_{t+1|t+1} - \hat{\underline{h}}_{t+1}(T)] \tag{12}$$

A gradient type of algorithm for solving this problem would have the following form:

$$\underline{\hat{T}}^{(\ell+1)} = \underline{\hat{T}}^{(\ell)} + K \left[\frac{\partial \underline{\hat{h}}_{t+1}[\hat{T}^{(\ell)}]}{\partial \underline{T}}\right]' \left[\underline{\hat{h}}_{t+1|t+1} - \underline{\hat{h}}_{t+1}[\hat{T}^{(\ell)}]\right] \quad (13)$$

The matrix $\partial \underline{\hat{h}}_{t+1}[\hat{T}^{(\ell)}]/\partial \underline{T}$ is called the *sensitivity matrix*, and is derived by Kitanidis [4]. The matrix K is an arbitrary matrix of step sizes. We replace ℓ with $\ell + 1$ and insert $\underline{\hat{T}}^{(\ell+1)}$ into Equation (6), which gives a new estimate of heads to place in Equation (13). This continues until convergence. The final convergent $\underline{T}$ estimates are designated $\underline{\hat{T}}_{t+1|t+1}$, and we are ready to proceed to period t + 1.

The primary advantage of this approach over the more complicated extended Kalman filter approach as presented by Kitanidis [4] is that the final transmissivity estimates $\underline{\hat{T}}_{t+1|t+1}$ are consistent with the head estimates $\underline{\hat{h}}_{t+1|t+1}$. Also, there is no need to attempt to estimate transmissivity covariances. This information is not needed for the optimal development algorithm, described in a subsequent section. Our understanding of the nature of the uncertainty of the system is totally embodied in $P_{h(t+1|t+1)}$.

C. Forecast Mode

As will be elaborated subsequently, it is desirable to be able to forecast future piezometric head levels under various assumed well field developmental patterns. We can also use a portion of the Kalman filter algorithm for generating such forecasts.

In the forecast mode, we assume that current estimates of transmissivity $\hat{T}_\tau$ at the beginning of current real-time period τ stay constant, since we have no new observations with which to update them. However, if the C matrix can be influenced by future decisions [i.e., we have $C(\underline{u}_t)$], then we must include (10) in updating the covariances. This is of course true for our case, since decisions to drill wells also mean an augmentation of the data collection network, which therefore influences observation matrix C. Therefore, letting $P_\tau \stackrel{\Delta}{=} P_{h(\tau|\tau)}$ and combining (7), (10), and (11):

$$\underline{\hat{h}}_{t+1} = \phi(\underline{\hat{T}}_t)\hat{h}_t + \psi(\underline{\hat{T}}_t)\underline{u}_t + \Omega(\underline{\hat{T}}_t)\underline{c}_t \quad (14)$$

$$P_{t+1} = [\phi(\underline{\hat{T}}_t)\ P_t\ \phi(\underline{\hat{T}}_t)' + Q] - G_h C(\underline{u}_i)[\phi(\underline{\hat{T}}_t)\ P_t\ \phi(\underline{\hat{T}}_t)' + Q] \quad (15)$$

(for $t = \tau,\ldots,N$)

where G_h is given by (11). Notice that in the forecast mode, the covariances tend to increase with time, rather than decrease when observations are available, indicating the greater uncertainty associated with forecasts further into the future.

D. Estimating Noise Covariances

The noise covariance matrices Q and R must be estimated, along with initial $P_{h(1)}$. This is largely a subjective matter. As pointed out by Mehra [27], one advantage of the Kalman filter is that

> . . . the forecaster can use his judgement regarding the relative accuracy of the model values for observations to select appropriate values for noise covariance matrices Q and R. He can then examine the actual operation of the filter and adjust these values on-line if the situation changes at a later time.

Even though it is possible to obtain accurate measurements of water level in a well, the measurement location may not coincide with the node point in a grid and must be extrapolated. There are other factors, such as *well loss*, which make it possible that the level observed in the well does not reflect the actual piezometric head immediately surrounding the well.

The matrix $P_{h(1)}$ can be started as a diagonal matrix with large elements (e.g., 10^4 to 10^6). As observations are obtained in real time, these covariances will reduce, reflecting the decreasing uncertainty as to the system state. If these variances are initially too small, the model will tend to weight model estimates too heavily over subsequent data that may indicate different estimates. On the other hand, if the variances are set too high, convergence may be extremely slow. The great flexibility in selecting noise and error covariances is therefore both an advantage and a disadvantage. Often a large amount of guesswork is needed.

McLaughlin [5] has pointed out some of the characteristics of groundwater systems that tend to militate against successful application of Kalman filtering, such as the length of the time intervals involved. Kalman filtering is based on statistical theory, which in turn is based on the law of large numbers. It may not be possible to take enough data (both in a spatial and temporal sense) within a reasonable amount of time to assume any statistical significance of the head and transmissivity estimates. In addition, obtaining data is expensive. Though there is a variety of geophysical and other techniques for estimating basin characteristics, the most reliable is *pump-testing*. This usually requires both a pumping well for producing the excitation (and hopefully obtaining water supply also) and an observation well for measuring the response. Such wells are expensive, so there is little room for error in selecting the best locations. In addition to these problems, the potentially large number of spatially distributed groundwater basin state variables presents a challenging computational problem.

Other difficulties have been alluded to previously, such as uncertainties in boundary conditions. In spite of all these problems, the authors believe that the Kalman filter is worth trying for groundwater basin development and management problems. As will be discussed subsequently, the algorithm can be used as a tool for the hydrogeologist responsible for developmental decisions. If the algorithm produces estimates that are counterintuitive to him, he can always reject them and rely solely on his own experience and intuition. But the algorithm is flexible and adaptable, and can be updated and modified if new information warrants.

VI. OVERVIEW OF DECISION FRAMEWORK

The previous section has presented the Kalman filter algorithm for successively estimating the parameters and state of the groundwater basin as new data become available. Both first and second order statistical information is generated, with the latter being particularly useful for risk analysis. The purpose of this section is to propose a methodology whereby the Kalman filter algorithm is explicitly incorporated into a dynamic programming algorithm for finding optimal sequential decisions for groundwater basin development. Dynamic programming is selected over other mathematical programming techniques, as well as optimal control theoretic approaches, because of the potential nonconvexity of our optimization problem and the existence of important state-space constraints. The nonconvexity can arise from concave capital investment costs, as well as operational costs as a multiplicative function of discharge and pumping lift or head.

Though dynamic programming is an effective approach to solving sequential decision problems with nonlinear, nonconvex cost functions and state-space constraints (i.e., as pointed out by Nemhauser [28], dynamic programming is actually enhanced by the existence of state-space constraints), the so-called *curse of dimensionality* looms as a problem. A modification of dynamic programming is presented in [3] which circumvents this problem.

Figure 3 gives a schematic of a proposed decision-making framework. The process begins with specification of budgetary limitations by the groundwater user and his projected water supply requirements. The hydrogeologist, based on his experience, judgement, and intuition, makes initial decisions about likely locations in the basin for development. He should specify several more locations than actually needed so as not to trivialize the decision algorithm. Any initially available groundwater basin data are then utilized to develop a finite difference groundwater basin model. This may require geophysical test information and test-hole drilling and analysis. Initial estimates of the parameters, boundary

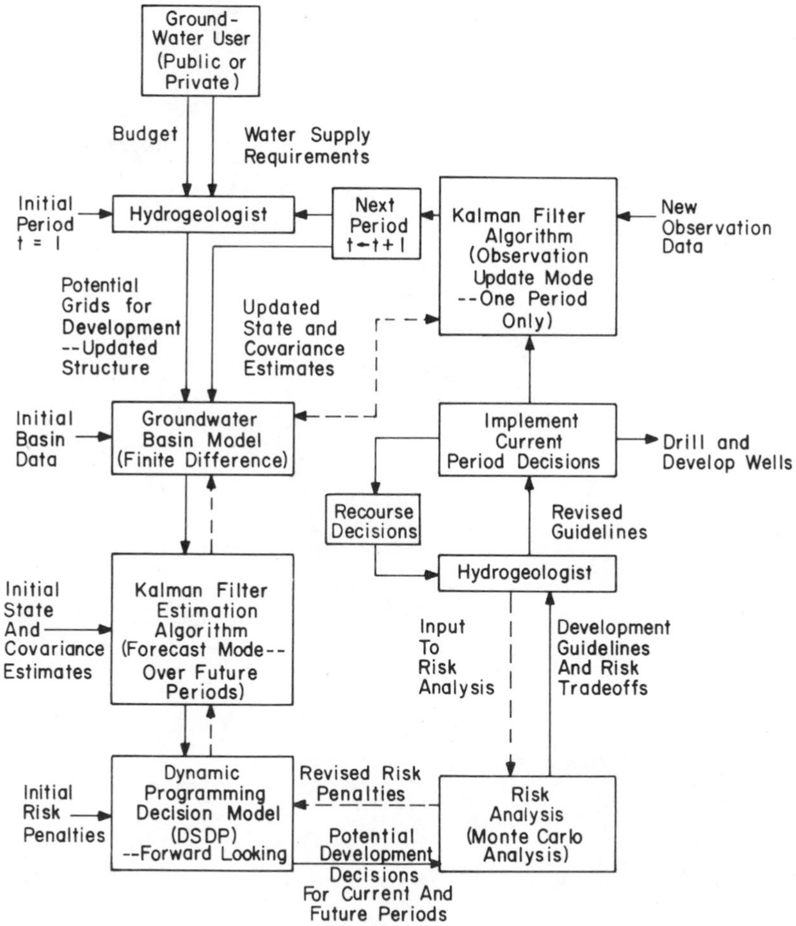

Figure 3 Overview of the sequential decision process.

conditions, initial conditions, grid structure, nonhomogeneities, stratification, etc., are made, but all of these can be updated as new information becomes available.

The Kalman filtering algorithm previously described is then applied in conjunction with the groundwater basin model to generate predictions of future water levels and pressure head distributions under a variety of potential developmental patterns. The forecast mode of the Kalman filter model is used here since no new observations are available. Also, the current basin parameter estimates are used for all future head predictions. The developmental patterns are screened via the dynamic programming model, which is referred to as decision-space dynamic programming (DSDP) [3]. Several iterations between the Kalman filter model and the DSDP algorithm are required.

Risk is dealt with indirectly. The covariance matrix is used as a penalty term in such a way that decisions resulting in heads close to minimum depths associated with *well pump bowls*,* along with high error covariances for these heads, are penalized more than safer decisions. To be feasible, a decision set must be such that the wells drilled are *expected* to meet the current and future water supply requirements. Through use of the indirect penalty approach, many of these will be rejected since they are too risky. That is, the chance of wells going dry and the water supply failing to be met may be great for these cases.

Since the penalty function approach is only indirect and does not explicitly quantify risk, it may be advisable to use the current estimated statistics of the future heads associated with a given decision policy to synthetically generate future realizations via Monte Carlo analysis and test the viability of the proposed decision. If failure occurs too often, the dynamic programming decision model can be run again with a higher penalty dispersion factor multiplied by the state error covariance matrix, and the above process repeated. The hydrogeologist has an important interaction here with respect to analyzing risk tradeoffs. An excessive number of iterations of this process may not be required since interpolations between extreme risk levels can be used.

Implementation of the *current period* development guidelines is next, subject to any revisions deemed necessary by the hydrogeologist. The dy-

*Structure into which water is drawn, pressurized, and ejected.

namic programming model will of course give current as well as future decision guidelines, but it is only the current ones that are actually implemented. New information obtained from the current period conditions may totally change the context for decision-making in the following periods.

As the hydrogeologist-revised guidelines are implemented and well test information begins to be collected, unforeseen conditions may appear which require rapid recourse decisions by the hydrogeologist in order to make sure that the water supply requirement is met. This may mean nonoptimal drilling based almost entirely on the hydrogeologists's intuition.

Once all drilling is completed for the current period and new well level data are collected, we can employ the full Kalman filter algorithm in conjunction with the groundwater basin model. Well observation data collected over the current period are utilized to update the head estimates, parameter estimates, and head error covariances. We then move to the next period. The hydrogeologist at this point again proposes potential areas for well drilling based on the updated model information and his own judgement, and the entire decision process begins again. It should be noted that the groundwater basin model could at this point be totally restructured if evidence from current data suggests the existence of various porous media formations.

It should be emphasized that this decision framework is primarily applicable to developing interfering well fields. It would be less applicable to areas where wells are drilled far apart since there is less extrapolatable information upon which to base new drilling decisions. The emphasis is on a decision framework for drilling pumping wells rather than observation wells and test holes. It is assumed here that these decisions are made outside this particular decision framework, although future work should focus on properly integrating all these aspects of the problem.

It is also assumed that the groundwater user has a legal right to the water supply requirements he has projected. Ideally, the model scope should be extended both spatially (i.e., to include adjacent water users who could be adversely affected by development of the basin under study) and temporally. This includes both groundwater and adjacent surface water users, such as downstream users who could be injured by stream depletions caused by groundwater pumping.

VII. COMPUTATIONAL EXPERIENCE

A hypothetical example was used to demonstrate the Kalman filter model and the dynamic programming algorithm. These results are not complete and much more work is needed, but they do provide some insight into the computational feasibility of the proposed methodology.

The geometry of the example aquifer we are using is the same as that shown in Figure 2. The assumptions and data associated with this example are listed as follows.

1. The boundary nodes are assumed to be impermeable, except for a constant head boundary on the right side. This could be a pond, lake, stream, or even another aquifer separated by a leaky layer. A pumping well is located in the center grid of Figure 2, and is discharging at a rate of 2000 m^3/day.
2. Each grid is 100 meters × 100 meters in size.
3. The aquifer is confined, with an assumed storage coefficient S = 0.01 for each grid. Our computer program is currently written for homogeneous values of S over the aquifer, but could be easily modified for varying S values.
4. The bedrock elevation is uniformly 10 m above an arbitrary datum. The top of the confined aquifer is 20 m in elevation.
5. Initial water levels are 70 m above the datum.
6. At each of the nine interior grids, inexpensive (i.e., relative to a pumping well) observation wells measure the heads at each time step.
7. Permeability of the aquifer is initially estimated to be 1.0 m/day for all links, with the *true* value at 1.8 m/day. Since the aquifer is 10 m thick, the initial transmissivity estimates are 10 m^2/day. We could have just as easily assumed the aquifer to be nonhomogeneous, but decided to use a homogeneous case as a first step. This in no way detracts from the generality of the algorithm to consider nonhomogeneous T values.
8. The error statistics were specified as follows:
   (a) Initial head error covariance matrix $P_{h(0)} = 0.25 \times I$ (where I is a 9 × 9 identity matrix, correcponsing to the nine interior grids)
   (b) The model noise:
   $Q = 0.5 \times I$
   (c) The observation noise:
   $R = 0.1 \times I$

These noise covariances are smaller than what would probably be selected if the algorithm were applied to a real case. Considerable experimentation is needed to determine the noise characteristics that best facilitate convergence of the filter. Our experimentation with these values is limited as of this writing.

9. Observations of the system state are made at the following times (in days):

0.01	0.02	0.05	0.12	0.26	0.60	
1.2	2.6	4.0	5.5	7.0	8.5	10.0

The extended Kalman filter algorithm proposed by Wilson, et al. [7] was programmed and run for this example problem. This approach requires estimates of error covariances for transmissivity and cross covariances between head and transmissivity. These are difficult to estimate, and are not needed for the algorithm we have presented. In addition, the Wilson approach generates estimates of T which may be completely inconsistent with heads predicted by the groundwater model. Our approach guarantees this consistency. We found that the Wilson approach generated extremely unrealistic transmissivity estimates (even negative values) unless the elements of the noise covariance matrices (Q and R) were set to unrealistically low levels. Indeed, Wilson, et al. [7] report that they set Q = [0] and $R = 8.33 \times 10^{-6}$ I for their example program. This means they were assuming the model to be error-free and water level observations to be highly accurate.

Some convergence results for the approach presented herein are given in Tables 1 and 2. Table 1 shows convergence of the $\underline{T}$ values (at 7.5 days) for the optimization problem of Equation (12). The gradient-type algorithm of Equation (13) was modified somewhat because of certain instabilities that were discovered. It was found that the elements of the sensitivity matrix varied by orders of magnitude, depending on the current period. This produced wide variation in the step sizes taken by the gradient algorithm. Instead of trying to use the step size matrix K to compensate for this (which would have required allowing it to vary with time) we decided to use the gradient to give the proper direction of search only. An initial step size of 25% of the current T estimates was used. Whenever the gradient indicated a reversal in search direction for a particular T component, the step size was quartered. Obviously, there is much room for improvement in solving this optimization problem.

Table 1 Convergence Characteristics of the Optimization Problem of Equations (12) and (13) for a Selected Time Period (@ 7.5 days)

Iteration No.	TX(1,3) (m^2/day)	TX(3,3) (m^2/day)	Sum of Squares Fitting Error
0	21.25	19.38	5.99
1	20.62	19.22	5.90
2	20.0	19.06	5.05
3	19.37	18.91	4.56
4	19.69	18.75	4.42
5	19.53	18.59	4.18
6	19.38	18.44	4.05
7	19.22	18.28	3.84
8	19.06	18.13	3.68
9	18.91	17.97	3.52
10[a]	18.75	17.81	

[a]Note: The iterative process is designed to terminate when the estimated heads are within a certain percentage of the true heads; in this case, 2.5%.

It should be noted that we actually used two models for predicting groundwater heads. The model of Equation (17) was only used for providing transition matrices for Equation (18). An implicit finite difference model developed by McWhorter and Sunada* was used for predicting heads $\hat{\underline{h}}_{t+1}$. The discrepancy between the two models is not believed to be significant.

Convergence results for the Kalman filter are presented in Table 2, with a portion of the results shown graphically in Figure 5. Convergence of the $\underline{T}$ values is quite rapid until the percentage error is less than 10%; at which point it is rather unstable, due to the assumed noise characteristics of the system. The convergence of the variance estimates is not very interesting for this example. We have found that the filter is quite sensitive to the relative magnitudes of model noise versus observation noise covariances, as well as their absolute magnitudes. Much more experimentation is needed on behavior of the filter with changes in the assumed noise characteristics.

---

*User's Manual for Program GRWATER, Appendix C of *Ground-Water Hydrology and Hydraulics*, Water Resources Publications, Fort Collins, 1977.

Table 2 Convergence of the Kalman Filter for the Discharge Node (2,2) (See Figure 2)

Time Period	Time (Days)	True Trans-missivity (m^2/day)	Estimated TY(2,2) (m^2/day)	Estimated TX(2,3) (m^2/day)	True Heads (m)	Estimated Heads (m)	Variance (m^2)
0	0	18.0	10.0	10.0	70.0	70.0	0.250
1	0.01	18.0	12.5	12.5	69.8	70.1	0.088
2	0.02	18.0	15.0	15.0	69.6	70.0	0.085
3	0.05	18.0	17.5	17.5	69.0	69.1	0.085
4	0.12	18.0	20.0	16.9	67.7	68.0	0.085
5	0.26	18.0	22.5	19.4	65.4	65.6	0.085
6	0.6	18.0	21.9	18.8	60.7	60.6	0.085
7	1.2	18.0	21.3	21.3	54.6	54.8	0.085
8	2.6	18.0	20.0	20.0	46.0	45.9	0.084
9	4.0	18.0	19.4	19.4	40.2	40.2	0.084
10	5.5	18.0	18.8	18.8	35.7	36.2	0.084
11	7.0	18.0	17.5	17.5	32.0	32.2	0.084
12	8.5	18.0	16.3	16.3	28.9	28.7	0.084
13	10.0	18.0	17.5	17.5	26.2	26.1	0.084

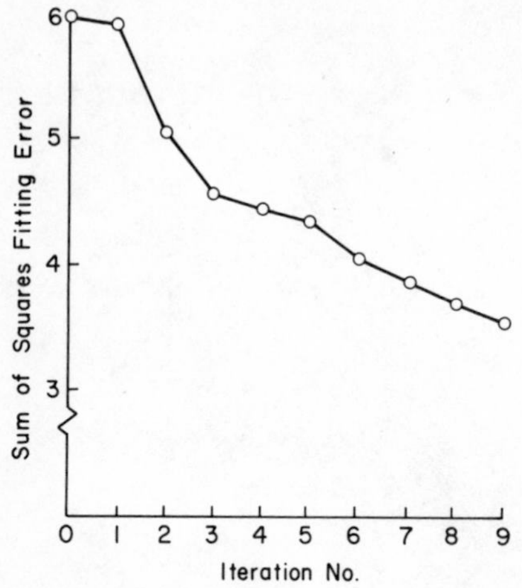

Figure 4 Convergence of fitting error.

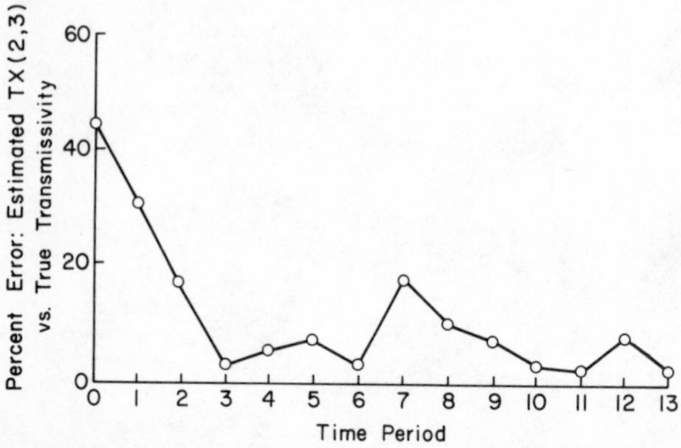

Figure 5 Convergence of percent error of TX(2,3) estimates.

Our experiments with the dynamic programming algorithm, using the Kalman filter in the forecast mode, are also quite limited. An example was run with five of the nine grids selected as potential locations for drilling pumping wells. Water supply requirements were specified such that one well would need to be drilled during each of five 100-day decision periods. The code was able to find the optimal decision sequence over the five periods in about 18 seconds on the CDC cyber 172 computer at Colorado State University. Only capital investment costs were considered in this example, and risk penalties were not included. Further work is continuing on inclusion of these factors.

ACKNOWLEDGMENTS

This research was partially supported by the National Science Foundation, Grant ENG-76-19814. The authors are grateful to Dr. Warren A. Hall, Department of Civil Engineering, and Dr. James P. Waltz, Department of Earth Resources, Colorado State University, for their valuable suggestions and guidance. The second author was supported in part by the National Wildlife Foundation.

REFERENCES

1. D. K. Todd, *Groundwater Hydrology*, New York: Wiley, 1959.
2. Y. Bachmat, B. Andrews, D. Holtz, and S. Sabastian, Utilization of numerical groundwater models for water research management, Grant No. R-803713, Robert S. Kerr Environmental Research Laboratory, Office of Research and Development, U.S. Environmental Protection Agency, Oklahoma, June 1978.
3. J. W. Labadie and D. R. Hampton, Dynamic programming under noisy state information: With application to water resources, presented at the Joint TIMS/ORSA National Meeting, Washington, D.C., May 5-7, 1980.
4. P. Kitanidis, A unified approach to the parameter estimation of groundwater models, M.S. Thesis, Massachusetts Institute of Technology, 1974.
5. D. B. McLaughlin, Potential applications of Kalman filtering concepts to groundwater basin management, in *Applications of Kalman Filter to Hydrology, Hydraulics, and Water Resources* (C.-L. Chiu, ed.), Stochastic Hydraulics Program, Department of Civil Engineering, University of Pittsburg, 1978.
6. R. A. Freeze and J. A. Cherry, *Groundwater*, Englewood Cliffs: Prentice-Hall, 1979.
7. J. Wilson, P. Kitanidis, and M. Dettinger, State and parameter estimation in groundwater models, in *Applications of Kalman Filter to Hydrology, Hydraulics, and Water Resources* (C.-L. Chiu, ed.), Stochastic Hydraulics Program, Department of Civil Engineering, University of Pittsburg, 1978.
8. C.-T. Chen, *Introduction to Linear System Theory*, New York: Holt, Rinehart and Winston, 1970.
9. M. S. Hantush, Economical spacing of interfering wells, in *Groundwater in Arid Zones*, 350-364, International Association of Scientific Hydrology, Publication 57, 1961.
10. W. M. Alley, E. Aguado, and I. Remson, Aquifer management under transient and steady-state conditions, *Water Research Bulletin 12*(October 1976), 963-972.
11. Hydraulics and economics of well field layout, *Public Works 108*(January 1977), 40-41.
12. J. R. Mount, A simplified technique for well-field design, *Groundwater 7*(May/June 1969), 5-8.

13. H. O. Pfannkuch and B. A. Labno, Design and optimization of groundwater monitoring networks for pollution studies, *Groundwater 14*(November/December 1976).

14. E. Bostock, E. S. Simpson, and T. G. Roefs, Minimizing costs in well field design in relation to aquifer models, *Water Resources Research 13*(April 1977).

15. S. Chang and W. W.-G. Yeh, A proposed algorithm for the solution of the large-scale inverse problem in groundwater, *Water Resources Research 12*(June 1976), 365.

16. R. L. Lenton, J. C. Schaake, Jr., and I. Rodriguez-Iturbe, Potential application of Bayesian techniques for parameter estimation with limited data, Department of Civil Engineering, M.I.T., Cambridge, Massachusetts, 1973.

17. J. S. Gates and C. C. Kisiel, Worth of additional data to a digital computer model of a groundwater basin, *Water Resources Research 10* (October 1974).

18. J. P. Delhomme, Kriging in hydrosciences, Ph.D. Dissertation, Centre d'Information Géologique, Ecôle Nationale Superieure des Mines de Paris (University of Paris), 1976.

19. L. W. Gelhar, Effects of hydraulic conductivity variations on groundwater flows, Second International Symposium on Stochastic Hydraulics, Lund, Sweden, August 2-4, 1976.

20. R. E. Kalman, A new approach to linear filtering and prediction problems, *J. Basic Engineering 82*(1960), 35-45 (trans. A.S.M.E.).

21. D. P. Lettenmaier and S. J. Burges, Use of state estimation techniques in water resource system modeling, *Water Resources Bulletin 12* (February 1976).

22. K. J. Åström, *Introduction to Stochastic Control Theory*, New York: Academic Press, 1970.

23. P. Young, Recursive approaches to time series analysis, Institute of Mathematics and Its Applications--Bulletin, Vol. 10, Nos. 4 and 5 (1974).

24. J. W. Labadie, A surrogate-parameter approach to modeling groundwater basins, *Water Resources Bulletin 11*(February 1975).

25. L. W. Nelson and E. Stear, The simultaneous on-line estimation of parameters and states in linear systems, *IEEE Trans. Automatic Control AC-21*(February 1976).

26. D. Graupe, *Identification of Systems*, New York: Van Nostrand Reinhold, 1972.

27. R. K. Mehra, Practical aspects of designing Kalman filters, in *Applications of Kalman Filter to Hydrology, Hydraulics, and Water Resources* (C.-L. Chiu, ed.), Stochastic Hydraulics Program, Department of Civil Engineering, University of Pittsburg, 1978.

28. G. L. Nemhauser, *Introduction to Dynamic Programming*, New York: Wiley, 1966.

INDEX